Livestock Emergency Guidelines and Standards

Third edition

LEGS

Practical Action Publishing Ltd
25 Albert Street, Rugby, Warwickshire, CV21 2SG, UK

www.practicalactionpublishing.com

© Livestock Emergency Guidelines and Standards – LEGS, 2023
First published 2009. Second edition 2014. Third edition 2023

The moral right of the editors to be identified as editors of the work and the contributors to be identified as contributors of this work have been asserted under sections 77 and 78 of the Copyright Design and Patents Act 1988.

All rights reserved. No part of this publication may be reprinted or reproduced or utilised in any form or by any electronic, mechanical, or other means, now known or hereafter invented, including photocopying and recording, or in any information storage or retrieval system, without the written permission of the publishers.

Product or corporate names may be trademarks or registered trademarks, and are used only for identification and explanation without intent to infringe.

A catalogue record for this book is available from the British Library.

ISBN 978-1-78853-246-4 Paperback
ISBN 978-1-78853-248-8 Electronic book

Citation: LEGS (2023) Livestock Emergency Guidelines and Standards, 3rd edition. Rugby, UK: Practical Action Publishing. http://doi.org/10.3362/9781788532488

Since 1974, Practical Action Publishing has published and disseminated books and information in support of international development work throughout the world. Practical Action Publishing is a trading name of Practical Action Publishing Ltd (Company Reg. No. 1159018), the wholly owned publishing company of Practical Action. Practical Action Publishing trades only in support of its parent charity objectives and any profits are covenanted back to Practical Action (Charity Reg. No. 247257, Group VAT Registration No. 880 9924 76).

The views and opinions in this publication are those of the editors and do not represent those of Practical Action Publishing Ltd or its parent charity Practical Action. Reasonable efforts have been made to publish reliable data and information, but the editors and publisher cannot assume responsibility for the validity of all materials or for the consequences of their use.

Edited by Helen de Jode and Cathy Watson
Design by www.truedesign.co.uk
Typesetting by River Valley Technologies

www.livestock-emergency.net

Contents

LEGS framework

How to use this handbook
p.12

Chapter 1:
Introduction to LEGS
p.18

Chapter 2:
LEGS Principles
p.46

Chapter 3:
Emergency response planning
p.76

Technical standards

Chapter 4:
Technical standards for livestock feed
p.130

Chapter 5:
Technical standards for the provision of water
p.168

Chapter 6:
Technical standards for veterinary support
p.204

Chapter 7:
Technical standards for livestock shelter and settlement
p.248

Chapter 8:
Technical standards for livestock offtake
p.288

Chapter 9:
Technical standards for the provision of livestock
p.320

Annexes
p.362

Index
p.379

Contents

How to use this handbook

14	What is the purpose of the LEGS Handbook?
14	Who should use the Livestock Emergency Guidelines and Standards?
15	What does the LEGS Handbook contain?
15	What difference will using the LEGS Handbook make?
16	What other resources does LEGS provide?

Chapter 1: Introduction to LEGS

20	What is the basis of LEGS?
23	Which livestock, livestock keepers and assets does LEGS cover?
26	What are the LEGS livelihoods objectives?
28	How does LEGS define emergencies and what are their impacts?
32	What other key concepts and issues are relevant to LEGS?
37	How does LEGS relate to Sphere and the Humanitarian Standards Partnership?
39	What sources of information exist for understanding more about LEGS?
43	What further information is available on topics that LEGS does not cover?

Chapter 2: LEGS Principles

48	Introduction
48	Principle 1: Supporting livelihoods-based programming
52	Principle 2: Ensuring community participation
55	Principle 3: Responding to climate change and protecting the environment
57	Principle 4: Supporting preparedness and early action
59	Principle 5: Ensuring coordinated responses
61	Principle 6: Supporting gender-sensitive programming
63	Principle 7: Supporting local ownership
66	Principle 8: Committing to monitoring, evaluation, accountability and learning (MEAL)

69	Appendix 2.1: LEGS alignment with the Sphere Protection Principles
70	Appendix 2.2: How LEGS Principles support the Core Humanitarian Standard Commitments
73	References and further reading

Chapter 3: Emergency response planning

78	Introduction
79	Stage 1: Initial assessment
87	Stage 2: Response identification
90	Stage 3: Analysis of technical interventions and options
96	Stage 4: Response plan
98	Monitoring, evaluation, accountability and learning (MEAL) guidance for LEGS interventions
105	Appendix 3.1: Initial assessment – suggested participatory methods
109	Appendix 3.2: Initial assessment – example checklist (partially completed)
111	Appendix 3.3: Initial assessment – data collection template
112	Appendix 3.4: PRIM – templates
114	Appendix 3.5: PRIM – samples
119	Appendix 3.6: Cash and voucher assistance – response modalities, delivery mechanisms and decision tree
124	Appendix 3.7: Response plan – template
125	Appendix 3.8: Response plan – example of livestock offtake intervention
126	References and further reading

Contents continued

Chapter 4: Technical standards for livestock feed
- **132** Introduction
- **135** Options for ensuring feed supplies
- **141** Timing of interventions
- **142** Links to other LEGS chapters and other HSP standards
- **143** LEGS Principles and other issues to consider
- **148** Decision tree for livestock feed options
- **151** The standards
- **161** Appendix 4.1: Assessment checklist for feed provision
- **163** Appendix 4.2: Examples of monitoring and evaluation indicators for livestock feed interventions
- **165** References and further reading

Chapter 5: Technical standards for the provision of water
- **170** Introduction
- **173** Options for water provision
- **177** Timing of interventions
- **178** Links to other LEGS chapters and other HSP standards
- **179** LEGS Principles and other issues to consider
- **182** Decision tree for water provision options
- **186** The standards
- **199** Appendix 5.1: Assessment checklist for water points
- **201** Appendix 5.2: Examples of monitoring and evaluation indicators for water provision
- **202** Appendix 5.3: Considerations for water point management
- **203** References and further reading

Chapter 6: Technical standards for veterinary support

- **207** Introduction
- **208** Options for veterinary support
- **216** Timing of interventions
- **217** Links to other LEGS chapters and other HSP standards
- **218** LEGS Principles and other issues to consider
- **223** Decision tree for veterinary support options
- **226** The standards
- **241** Appendix 6.1: Assessment methods and checklist for veterinary support
- **243** Appendix 6.2: Examples of monitoring and evaluation indicators for veterinary support
- **245** References and further reading

Chapter 7: Technical standards for livestock shelter and settlement

- **251** Introduction
- **254** Options for shelter and settlement
- **259** Timing of interventions
- **260** Links to other LEGS chapters and other HSP standards
- **261** LEGS Principles and other issues to consider
- **265** Decision tree for livestock shelter and settlement options
- **268** The standards
- **282** Appendix 7.1: Assessment checklist for livestock shelter and settlement provision
- **284** Appendix 7.2: Examples of monitoring and evaluation indicators for livestock shelter and settlement
- **285** Appendix 7.3: Livestock shelter and climate challenges
- **287** References and further reading

Contents continued

Chapter 8: Technical standards for livestock offtake
- **290** Introduction
- **292** Options for livestock offtake
- **296** Timing of interventions
- **297** Links to other LEGS chapters and other HSP standards
- **298** LEGS Principles and other issues to consider
- **301** Decision tree for livestock offtake options
- **303** The standards
- **315** Appendix 8.1: Assessment checklist for livestock offtake
- **317** Appendix 8.2: Examples of monitoring and evaluation indicators for livestock offtake
- **319** References and further reading

Chapter 9: Technical standards for the provision of livestock
- **323** Introduction
- **325** Options for the provision of livestock
- **330** Timing of interventions
- **331** Links to the other LEGS chapters and other HSP standards
- **331** LEGS Principles and other issues to consider
- **338** Decision tree for provision of livestock options
- **341** The standards
- **352** Appendix 9.1: Assessment checklist for provision of livestock
- **354** Appendix 9.2: Examples of monitoring and evaluation indicators for the provision of livestock
- **358** Appendix 9.3: Discussion on minimum viable herd size
- **360** References and further reading

Annexes
364 Annex A: Glossary
374 Annex B: Abbreviations and acronyms
375 Annex C: Combined timing table
376 Annex D: Acknowledgements and contributors
379 Index

Praise for this book

'The revision of LEGS Handbook includes significant changes in content and form. The new structure, improved design, changes to the foundations chapters and Plain English used allow for easier access and understanding for both livestock specialists and humanitarians, even for first time users. The links to other chapters and other Humanitarian Standards Partnership (HSP) handbooks are also welcome.'

Maty Ba Diao, former Regional Coordinator, Projet Régional d'Appui au Pastoralisme au Sahel (PRAPS), at the Comité permanent inter-État de lutte contre la sécheresse au Sahel (CILSS)

'I welcome this third edition of the LEGS handbook, which is even more reader friendly, easy to use, relevant to changing contexts and supported by more case studies. It includes innovative tools to identify how to respond to emergencies, and has assessed past interventions and experiences to guide future planning. The consultative process and outreach have gone beyond the previous edition. I am sure the tools in the handbook will be very useful for both humanitarian and development actors working in the area of food security focusing on livestock-based livelihoods.'

Vikrant Mahjan, CEO, Sphere India

'This third edition has enriched content and more simplified language, and has been revised in line with community of practice comments as part of greater local ownership of the LEGS approach. The Handbook clearly illustrates the interlinked nature of lives and livelihoods, and the new chapter on emergency response planning integrating key tools will support the saving of lives and protecting the livelihoods of affected communities.'

Dr Kisa Juma Ngeiywa, former Director of Veterinary Services (DVS) Kenya

'Es importante para mí haber participado en la 3ª edición del Manual de LEGS, ya que es una fuente esencial de directrices para procedimientos orientados a preservar los medios de vida de la población más vulnerable en eventos adversos. Me brindó también la oportunidad de formar parte de un selecto grupo de profesionales del sector a nivel mundial que atienden con humanidad a los animales.' [Contributing to the 3rd edition of the LEGS Handbook was important to me. It is an essential source of guidelines for procedures aimed at preserving the livelihoods of at-risk people in emergencies, and it also provided a unique opportunity to be among a select group of world professionals with a humane approach to livestock issues.]
Norman Ernesto Mora Cerda, Director de Capacitación y Respuesta, Codirecciones del SINAPRED, Nicaragua

'LEGS is the reference guide for implementing quality livestock interventions in emergency settings that strengthen the recovery capacity and overall resilience of affected populations. This third edition of the LEGS handbook is more user-friendly, includes new core principles such as ensuring community participation or supporting gender-sensitive programming, and offers updated standards, key actions and guidance notes. As one of the contributors to the LEGS handbook since its inception, VSF is committed to continuously training its staff to ensure a thorough understanding of the LEGS principles and standards and their application. The handbook is thus an essential complement to the trainings, and this improved third edition will be an easy-to-use manual for field staff at any stage of a crisis response.'
Margherita Gomarasca, Coordinator, VSF International

'The LEGS handbook provides a common language and framework for livestock practitioners, enabling joint planning and clear mapping of needs and gaps. The new edition will be instrumental to enable teams to undertake pre-emptive actions to reduce the loss of livestock assets and production with its key feature on preparedness and early action. Furthermore, the inclusion of additional details on cash and voucher assistance will strengthen cash plus livestock interventions. The guidelines have been, and will continue to be, instrumental in quality responses through continuous learning and exchange amongst partners, including policy makers.'
Rein Paulsen, Director of Emergencies and Resilience Office, Food and Agriculture Organization of the United Nations

LEGS framework
How to use this handbook

How to use this handbook

- 14   What is the purpose of the LEGS Handbook?
- 14   Who should use the Livestock Emergency Guidelines and Standards?
- 15   What does the LEGS Handbook contain?
- 15   What difference will using the LEGS Handbook make?
- 16   What other resources does LEGS provide?

Photo © FAO/Michael Tewelde

How to use this handbook

What is the purpose of the LEGS Handbook?

The Livestock Emergency Guidelines and Standards (LEGS) Handbook provides minimum standards and guidelines for use in humanitarian emergencies that impact livestock. The handbook helps support the livelihoods of livestock-keeping communities affected by emergencies. It prioritises those living in lower- and middle-income countries and aims to improve the quality of emergency humanitarian response.

The LEGS Handbook includes practical decision-making tools for planning a livestock emergency response. It provides standards, key actions and guidance notes for undertaking emergency interventions in six technical areas: livestock feed, water, veterinary support, shelter, livestock offtake, and the provision of livestock.

The LEGS Handbook highlights the importance of protecting livestock-based livelihoods during emergencies, and of rebuilding livelihoods after emergencies. Livelihoods-based programming is the first of eight LEGS Principles that users of the handbook are asked to follow.

All LEGS guidance is based on technical expertise and evidence-based good practice.

Who should use the Livestock Emergency Guidelines and Standards?

LEGS mainly targets those who provide emergency assistance in areas where livestock make an important contribution to livelihoods. These may include people based in affected communities; those working for national or international NGOs; those working in local, regional or national government; people working for bilateral or multilateral agencies; or private sector operators. Users of the handbook will also be the emergency-affected communities themselves. The handbook uses the term agencies to cover all who respond to emergencies impacting livestock.

LEGS guidance clearly explains how and why emergencies, livestock and livelihoods overlap. The handbook will therefore be useful to both livestock specialists and humanitarians involved in emergency preparedness, response and recovery. The handbook is also relevant for donors, policy makers and others whose funding, actions and policies influence response interventions. It may also be useful for educational institutions.

What does the LEGS Handbook contain?

The LEGS Handbook contains three framework chapters. It is important to read these first. *Chapter 1* introduces LEGS, including its foundations in Sphere and links with the Humanitarian Standards Partnership (HSP). *Chapter 2* explains the eight principles that underpin LEGS. *Chapter 3* provides the practical tools needed for planning a livestock-based emergency response. These framework chapters are the basis for the technical standards chapters that follow.

The six technical standards chapters *(Chapters 4, 5, 6, 7, 8, 9)* all share a common structure. They should be used for emergency response planning and as an ongoing reference source. They first identify why each specific technical intervention is important for supporting livestock and livelihoods. They then outline the benefits and challenges of the potential response options, and in which phase of a humanitarian emergency they are most applicable. The standards sections then set out:

- Standards – qualitative statements of the minimum to be achieved in any emergency in any context;
- Key actions – practical steps or actions for achieving the LEGS Standards, not all of which may be relevant to all situations;
- Guidance notes – to be read in conjunction with the key actions. They explain particular issues and how to address any practical difficulties when applying the LEGS Standards.

A glossary is provided at the end of the handbook containing important terminology. Glossary words are highlighted in colour the first time they appear.

LEGS framework
How to use this handbook

What difference will using the LEGS Handbook make?

The LEGS Handbook explains the critical importance of livestock assets for many at-risk communities. It aims to ensure that livestock-based livelihoods are not destroyed by well-meaning but inappropriate interventions during humanitarian emergencies. Use of the tools in the LEGS Handbook helps ensure livestock emergency responses are selected, designed and implemented effectively. The most appropriate, feasible and timely livestock interventions can be identified from the six LEGS technical intervention areas. The eight LEGS Principles provide a guidance structure that is applicable in all contexts, helping users to make appropriate decisions and apply the standards effectively.

What other resources does LEGS provide?

This third edition of the LEGS Handbook is available in hard copy format; on the Humanitarian Standards Partnership Interactive Handbook site; and as a downloadable pdf on the LEGS website (see www.livestock-emergency.net), where there is also further information about LEGS: resources include briefing papers, further tools and case studies. Some case studies cover the process: how the interventions were carried out, in other words. Others describe their impact: the effect on the livelihoods of livestock keepers. There are also details about the LEGS training programme and how to sign up to the LEGS e-newsletter.

LEGS framework
How to use this handbook

Chapter 1: Introduction to LEGS

20   What is the basis of LEGS?

23   Which livestock, livestock keepers and assets does LEGS cover?

26   What are the LEGS livelihoods objectives?

28   How does LEGS define emergencies and what are their impacts?

32   What other key concepts and issues are relevant to LEGS?

37   How does LEGS relate to Sphere and the Humanitarian Standards Partnership?

39   What sources of information exist for understanding more about LEGS?

43   What further information is available on topics that LEGS does not cover?

Photo © Michael Benanav

>>> LEGS framework
Introduction to LEGS

Chapter 1: Introduction to LEGS

Chapter overview

This chapter introduces LEGS by answering the following questions:
- What is the basis of LEGS?
- Which livestock, livestock keepers and assets does LEGS cover?
- What are the LEGS livelihoods objectives?
- How does LEGS define emergencies, and what are their impacts?
- What other key concepts and issues are relevant to LEGS?
- How does LEGS relate to Sphere and the Humanitarian Standards Partnership?
- What sources of information exist for understanding more about LEGS?
- What further information is available on topics that LEGS does not cover?

What is the basis of LEGS?

Livelihood support

It is now well recognised that emergency response should not just be about saving human lives, but that humanitarian action must also consider maintaining the livelihoods of affected communities. Livelihoods are how people sustain themselves and their families, and are their means of making a living. Livelihoods are often defined as people's capabilities, assets, income and activities required for securing the necessities of life.

Livelihood support is the basis of LEGS, and this is reflected in the handbook in two ways. Livelihoods-based programming is the first of the LEGS Principles (see *Chapter 2*, as well as *Chapters 4, 5, 6, 7, 8, 9* for its relevance to each technical intervention). LEGS also follows a livelihoods-based approach in the design of emergency response plans *(see Chapter 3)*.

Where there is a crisis, LEGS supports the livelihoods of livestock-keeping communities. LEGS provides this support through planning and implementing response options from six technical intervention areas: feed, water, veterinary support, shelter, livestock offtake and the provision of livestock. These technical interventions target livestock assets and are aligned with one or more of the three LEGS livelihoods objectives (see below).

Evidence

LEGS draws on **evidence-based** good practice from around the world. As part of this evidence base, it tracks and uses evaluations, reviews and impact assessments of livestock and livestock-related interventions. This third edition of the LEGS Handbook is based on:

- editorial recommendations from five discussion papers on key topics, as well as feedback from the webinars held on each paper;
- a discussion paper on 'How to make LEGS more user-friendly';
- consultation workshops held in Mali, Kenya, the Philippines, India and Nicaragua;
- an online consultation on the second edition of the LEGS Handbook;
- the review and updating of the LEGS Impact Database (available on the LEGS website at www.livestock-emergency.net/legs-impact-database);
- a public consultation on the draft third edition.

This evidence is presented in the LEGS Handbook as:

1. references and useful further reading provided at the end of each chapter;
2. case studies of practical, worldwide examples – these are highlighted in each chapter and are available on the LEGS website (www.livestock-emergency.net/case-studies).

Human rights

LEGS follows a **rights-based approach** that aims to respect, protect and fulfil internationally agreed human rights. The principal international human rights legal framework is the Universal Declaration of Human Rights, including Article 1 *'All human beings are born free and equal in dignity and rights'*.

Article 25 is particularly relevant for LEGS as it encompasses the right to food. It states, *'Everyone has the right to a standard of living adequate for the health and well-being of himself and of his family, including food, clothing, housing and medical care…'*. (Note that this wording is from 1948, when masculine pronouns were commonly used as a default.)

Animal welfare

LEGS response options ensure good **animal welfare**. Since the two previous editions of the LEGS Handbook, international organisations have evolved their understanding of the importance of animal welfare. The World Organisation for Animal Health (WOAH) now actively promotes animal welfare.

LEGS links its technical interventions to one of the animal welfare assessment protocols that focus on animal-based indicators, the Five Animal Welfare Domains, as listed in *Box 1.1*. The LEGS technical chapters explain how each intervention connects with these animal welfare domains.

Box 1.1
Animal welfare: Five domains

1. **Nutrition** – factors that involve the animal's access to sufficient, balanced, varied, and clean food and water.
2. **Environment** – factors that enable comfort through temperature, substrate, space, air, odour, noise, and predictability.
3. **Health** – factors that enable good health through the absence of disease, injury, impairment with a good fitness level.
4. **Behaviour** – factors that provide varied, novel and engaging environmental challenges through sensory inputs, exploration, foraging, bonding, playing, retreating, and others.
5. **Mental state** – the mental state of the animal should benefit from predominantly positive states, such as pleasure, comfort, or vitality while reducing negative states such as fear, frustration, hunger, pain, or boredom.

Source: https://www.worldanimalprotection.us/blogs/five-domains-vs-five-freedoms-animal-welfare

---

LEGS Principles

LEGS is based on eight **principles.** These cover:

1. supporting livelihoods-based programming;
2. ensuring community participation;
3. responding to climate change and protecting the environment;
4. supporting preparedness and early action;
5. ensuring coordinated responses;

6. supporting gender-sensitive programming;
7. supporting local ownership;
8. committing to monitoring, evaluation, accountability and learning (MEAL).

Chapter 2 explains the LEGS Principles and their application, while each technical intervention chapter *(Chapters 4, 5, 6, 7, 8, 9)* explains how the interventions address or align with the eight LEGS Principles.

Which livestock, livestock keepers and assets does LEGS cover?

How LEGS defines livestock

Within LEGS, the term 'livestock' refers to animal species that support livelihoods and communities, particularly in lower- and middle-income countries. This means that LEGS focuses primarily on the animals that are kept by pastoralists or small-scale producers, and not by large-scale commercial livestock enterprises. These animals range from poultry, pigs, sheep and goats, to cattle, camels, water buffalo, llamas, yaks and donkeys, and any other animals that contribute to people's livelihoods. Animals that do not directly support livelihoods, such as companion animals, are not the primary concern of LEGS. However, it is recognised that the definition of 'companion animals' varies in different contexts. LEGS does not cover beekeeping or aquaculture either. These are addressed elsewhere *(see What further information is available on topics that LEGS does not cover?).*

Working animals

In countries impacted by humanitarian crises, equines (horses, mules and donkeys) are often fundamental to livelihoods. These working animals are sometimes critical for women's livelihoods, for transporting items to market or collecting water. Dogs may also be working animals in some contexts. When using LEGS for planning responses to emergencies, it is necessary to consider the role and importance that the animals have in the affected communities' livelihoods.

Emergencies affect working animals in specific ways. In drought situations, for example, donkeys, camels and llamas may work much more collecting water and supplies from increasingly distant locations. They therefore need extra feed to avoid progressively losing body condition in these circumstances. And, if a working animal dies, households are left even more at risk.

LEGS also specifically recognises the importance of livestock that work in response to humanitarian crises. Local working animals may be used for pulling or carrying items such as food aid during a drought. These animals often have experienced, or are experiencing, the impact of the crisis. As a result, they may be traumatised, injured or in poor condition.

Types of livestock keepers

Throughout the world, animals play significant roles in people's livelihoods in different ways and to varying degrees, as listed in *Box 1.2*.

Box 1.2
Examples of livestock keepers whose livestock contribute to their livelihoods

- Pastoralists, who are highly dependent on livestock
- Agro-pastoralists, who have a combination of herds and crops, but whose livestock are often the main livelihood asset
- Smallholder crop farmers who use cattle or buffaloes for farm power, where ploughing and manure are critical for crop production
- Smallholder livestock farmers who raise cows, goats, pigs or poultry, for example, and who may depend on working animals for transport; these livelihoods may rely heavily on the animals raised both as a source of nutrition and income
- Service providers, for example llama, mule or donkey cart owners, who may be totally dependent on the animals for their income
- Urban and peri-urban livestock owners, as urban animals can often contribute to livelihoods as a supplementary source of income or food (some of these livestock keepers, like urban fresh milk suppliers, are totally dependent on their animals; some urban women also depend on buying young goats and sheep and finishing them for market)

Many communities depend on both livestock and crops for their livelihoods. The Standards for Supporting Crop-related Livelihoods in Emergencies (SEADS) follows a similar format to LEGS and should be consulted alongside LEGS to support smallholder farming communities affected by humanitarian crises.

Other livelihoods, such as livestock traders and shopkeepers, feed suppliers, as well as animal health workers, depend on livestock and livestock keepers.

Livestock as livelihood assets

For communities who use livestock for their livelihoods, livestock are livelihood assets. LEGS defines livelihood assets as the resources, equipment, skills, strengths and relationships that are used by individuals and households to pursue their livelihoods.

Livelihood specialists categorise livelihood assets as social, human, natural, financial or physical assets. In livestock-dependent livelihoods, animals are like a 'living' bank account: financial assets that owners can sell for income when needed or save as financial capital. Additionally, in many communities, livestock are important social assets. This means they give status to their owner. They can also be exchanged or else gifted as part of social relations. Human livelihood assets include people's health, nutrition and capacity to work *(see Box 2.1 in Chapter 2: LEGS Principles)*. For many rural households, livestock play a key role in meeting household nutritional needs, and define their socio-economic role.

For livestock owners, the extent to which they are at risk from the effects of an emergency is strongly linked to their livestock assets. This is because the greater the value of their assets, the greater their capacity to cope with shocks. It is therefore essential to understand the role that livestock have in livelihoods, and the emergency's impact on these livelihood assets, to determine how appropriate a livestock-based response is.

Chapter 3: Emergency Response Planning describes how to use the LEGS tools when making an initial assessment to understand livestock assets and livelihoods. Non-livestock-focused interventions, such as crop-based support, food aid, and cash and voucher assistance (CVA) can also complement livestock-based responses. They help remove some of the pressure on livestock assets in the short term, making recovery more feasible.

On the other hand, there may be men and women who have survived for generations as livestock keepers but who are so affected by an emergency that their assets are severely depleted. For them, keeping livestock is no longer an option. Recurrent emergencies erode household capacity and enthusiasm for livestock keeping, as family members are dispersed, injured or killed, and labour and managerial expertise lost. Such households may prefer to be given assistance to establish themselves in alternative livelihoods.

LEGS framework
Introduction to LEGS

What are the LEGS livelihoods objectives?

LEGS emergency response is based on three livelihoods objectives.

LEGS technical interventions support crisis-affected communities to:

1. **obtain immediate benefits** using existing livestock assets; and/or
2. **protect key livestock assets**; and/or
3. **rebuild key livestock assets.**

When responding to emergencies affecting livestock-keeping communities, potential responses are assessed against these three LEGS livelihoods objectives. The chosen technical intervention (providing water, providing shelter, or any of the other interventions described in the technical chapters) should help to achieve at least one of the LEGS livelihoods objectives. The introduction in each technical chapter explains their links to the objectives. *Table 1.1* provides more explanation on how the objectives link to the technical interventions.

Table 1.1: The three LEGS livelihoods objectives and examples of interventions

LEGS livelihoods objective	Example LEGS interventions
1. To **obtain immediate benefits** using existing livestock assets. Providing rapid assistance based on existing livestock assets to generate income or food as immediate benefits.	**Livestock offtake:** The sale or slaughter of at-risk or unmarketable animals provides cash or food support to livestock keepers. This allows longer-term protection of remaining livestock assets.
	Feed and Water: Providing feed and water to productive animals may contribute to immediate improvements in household food security through livestock products.

continued over

LEGS livelihoods objective	Example LEGS interventions
2. To **protect** key livestock assets. Keeping livestock alive so that production can resume when the emergency is over.	**Veterinary support:** This can have a positive impact on protecting and rebuilding livestock assets at all stages of an emergency. It may take place in conjunction with other activities (for example, feed, water, provision of livestock) to increase asset protection. **Feed:** This is important for protecting remaining livestock assets during and after an emergency. In drought, this may be complemented by water provision. **Water:** This is important for protecting remaining livestock assets. **Shelter:** This responds to a range of livestock needs – protection against cold or hot climates; security; provision of a healthy environment for livestock and humans. It can involve preventive measures (for example, earthquake-resistant livestock shelters, or the use of elevated areas for flood protection), as well as measures to protect livestock assets after an emergency (for example, volcanic eruption).
3. To **rebuild** key livestock assets. Rebuilding assets when livestock losses have already occurred, and asset protection was not possible.	**Provision of livestock:** This may include helping livestock keepers to rebuild herds after an emergency. It may also involve the replacement of smaller numbers of animals, e.g. poultry, small ruminants and working animals, which contribute to livelihoods. It is appropriate in the recovery phase once the immediate aftermath is over and an assessment of asset loss is possible. **Feed/water/shelter/veterinary support:** See Objective 2. Continued intervention in the recovery phase is necessary to rebuild and strengthen livestock assets and reduce vulnerability to future emergencies.

How does LEGS define emergencies and what are their impacts?

LEGS and the disaster risk management cycle

Humanitarian crises and disasters are associated with substantial loss of human life relative to a normal situation, and/or considerable damage to physical infrastructure and assets. There are also many hazards and threats that do not necessarily cause a disaster.

While there is a range of terminology used to describe disasters and humanitarian crises, LEGS uses the term 'emergency' for situations in which a hazard or threat is beyond the capacity of the affected community to cope. Poorer communities tend to be less well prepared, less resilient, and have lower coping capacity. Other variables that influence the impact on a community are their level of exposure to the hazard and how at risk they are.

Because emergencies tend to recur, it is useful to think of them not as one-off events, but as part of a cycle. LEGS aligns with the disaster risk management (DRM) cycle, which considers risk prevention and reduction as well as reducing the losses from emergencies. LEGS interventions largely focus on response and recovery, while the DRM cycle (see *Figure 1.1*) also highlights the importance of identifying risks and preparing for an emergency prior to the actual onset. Because LEGS recognises the key role that preparedness plays in this cycle, this third edition of the LEGS Handbook includes preparedness standards in many of the technical chapters, underpinned by the LEGS preparedness principle (see *Principle 4: Preparedness*, in *Chapter 2*).

LEGS emergency types and phases

LEGS categorises emergencies into two main types:

1. slow-onset emergency;
2. rapid-onset emergency.

When an emergency involves a breakdown in national authority, for example because of conflict or political instability, it is classified as a complex emergency. In these cases, a slow-onset crisis such as a drought, or a rapid-onset event such as an earthquake, may compound these other factors. This may happen to the extent that large-scale assistance is required to meet the needs of significant numbers of affected people. When a crisis continues over time, it may become a protracted emergency, while in some regions cyclical natural hazards may lead to a chronic/recurring emergency.

Figure 1.1: The disaster risk management cycle

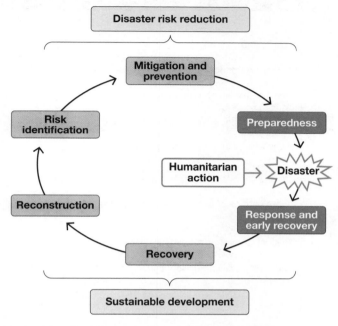

Source: Based on International Strategy for Disaster Risk Reduction (ISDR)

When planning an emergency response, it is important to be clear about what type of emergency it is. Examples of rapid-onset, slow-onset and complex emergencies are shown in *Table 1.2*.

Table 1.2: Emergency types and examples

Emergency type	Example emergencies
Slow-onset emergency	Drought
	Severe winter (for example, dzud in Mongolia)

continued over

LEGS framework
Introduction to LEGS

Emergency type	Example emergencies
Rapid-onset emergency	Flood
	Earthquake
	Volcanic eruption
	Storm (hurricane, typhoon, cyclone)
	Wildfires (naturally occurring or otherwise)
	Some animal or plant pests and diseases
Complex emergency	Short or long-lasting conflict with human displacement
	Combinations of conflict with another type of emergency, such as drought
	Political, economic and social crises
	Outbreak of transboundary animal disease or a pandemic superimposed on another type of emergency

To reflect how the different emergency types require interventions at different times, LEGS defines the phases within slow-onset and rapid-onset emergencies differently:

- For slow-onset emergencies, the LEGS phases are alert/alarm/emergency/recovery (based on the drought cycle management model).
- For rapid-onset emergencies, the phases are immediate aftermath/ early recovery/recovery.

Figure 1.2 illustrates how these phases correspond to the impact of the crisis.

The LEGS approach to emergency response planning *(see Chapter 3: Emergency response planning)* uses these different phases to help identify the most appropriate timing for the intervention that is being selected.

Figure 1.2: LEGS phases for slow-onset and rapid-onset emergencies

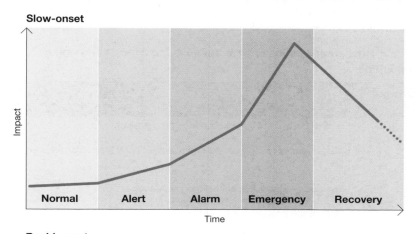

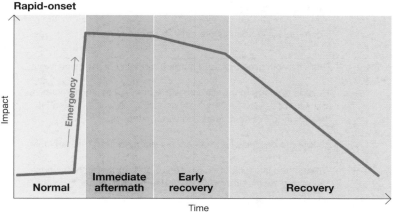

In a complex or protracted emergency, the phases can be determined according to whether the underlying emergency is slow- or rapid-onset. For example, in a complex emergency involving drought and civil war, the timing of interventions should align with the slow-onset phases. In a complex emergency resulting from an earthquake in a country affected by conflict, the rapid-onset phases may be the most appropriate for emergency planning.

LEGS framework
Introduction to LEGS

How do emergencies impact livestock keepers?

Emergencies impact communities that depend on livestock in many ways, as shown in *Table 1.3*.

Table 1.3: How emergencies impact animals and livestock-keeping communities

Emergency type	How it can affect livestock and communities
Slow-onset emergency	Animal body condition deteriorates progressively.
	Market prices for animals go down, while prices for human food go up.
	The cost of inputs like animal feed increases.
	Human food security and nutrition gradually worsen.
	As the emergency progresses, more animals die.
	If core breeding animals die, post-emergency recovery is hindered.
Rapid-onset emergency	There is rapid livestock (and human) mortality in the initial event.
	There is damage to infrastructure (like roads and livestock shelters) and services (such as veterinary services), as well as potential disruption to livestock feed and water supplies.
	People are displaced and perhaps separated from their animals.
	Where there is a lack of response support, further consequences can occur, such as more livestock deaths or disease.
Complex emergency	Access to grazing is reduced.
	Armed groups may steal animals.
	Conflict may prevent access to services or markets.
	Infrastructure, such as communication networks, is affected.
	People and animals are displaced.
	There are protracted and high levels of human food insecurity and malnutrition.

What other key concepts and issues are relevant to LEGS?

LEGS is consistent with several well-established and emerging concepts and issues, as outlined below.

Resilience analysis

Sustainable livelihoods and frameworks influenced development thinking in the last decade of the 20th century, and the principles of this approach continue to underpin LEGS. The concept of resilience builds on the sustainable livelihoods approach. It aims to integrate the impact of emergencies and the need to strengthen the ability to recover. Resilience applies equally at different levels: individual, household, community, local government, national government, and ecosystem. Different organisations define resilience in different ways depending on their mandate. A person's resilience depends on many factors that affect their ability to cope and adjust, such as their economic well-being, education, gender, health and age.

Strengthening communities' resilience helps them to protect themselves from the impact of future emergencies, with livestock playing a major role. For example, mixed farm households with crops and livestock produce their own fodder, reducing costs for feed/grain; they can also sell some animals for cash when crops fail. And pastoralists can accumulate animals in good years to protect them from shocks in bad ones. Resilient livestock systems also help improve food and nutrition outcomes.

The thinking behind the concept of resilience aligns with LEGS Principles and livelihoods objectives. However, there is no clearly documented evidence yet that emergency responses can have a positive impact on long-term resilience.

New pandemics and their impacts

A zoonotic disease, or zoonosis, is an infection or disease that is transmissible from animals to humans. Zoonoses that are a threat to human health from livestock are well recognised, including anthrax and rabies. More contacts at the human-animal interface, and interaction between wild species and farmed animals, can lead to increased risk of emerging infectious diseases. LEGS considers zoonoses within *Chapter 6: Veterinary support*, under the established practice of veterinary public health. Where people and animals are displaced, new cross-infection risks may arise (see *Chapter 7: Shelter*).

LEGS framework
Introduction to LEGS

Scientists have long recognised the risk of infectious virus spillover from livestock, with its potential for global human disease such as the potentially pandemic 'bird flu'. Whilst the pandemic due to the SARS-CoV-2 virus, causing Covid-19, was most probably a virus spillover event from a wildlife species, livestock keepers experienced significant consequences. The economic impacts and disruption to LEGS activities from Covid-19 are shown in *Table 1.4*.

Table 1.4: How the Covid-19 pandemic affected livestock livelihoods and LEGS activities

The pandemic's impact on livestock and livestock keepers	
Feed	In some places lockdown conditions caused feed shortages, for example, for dairy cows.
Movement of livestock	National and international movement restrictions (border closures) prevented pastoralists from moving their herds for grazing and trade.
Market closures	The lack of markets reduced livestock keepers' income.
	Transporting and selling livestock products (eggs, milk) became difficult.
	Led to shortages of livestock input supplies.
Veterinary services	In some places the pandemic restrictions disrupted government vaccination delivery.
The impacts of the pandemic on LEGS activities	
LEGS training	Training changed from face to face to online.
LEGS participatory analysis (initial assessment, response identification, MEAL)	Access to communities reduced or blocked.
	Care became necessary when using techniques that could lead to reduced social distancing.

Livestock and climate change: global debates

A substantial body of research and analysis related to global climate change shows that livestock are an important source of greenhouse gases. This finding is the basis for high-level policy debates on the future of livestock production and human food systems. A prominent policy position is that, globally, human food systems should become plant-based, with far lower, or zero, consumption of foods derived from animals. These debates influence international aid policies, including the extent to which livestock systems should be supported. The same debates also transfer across into humanitarian aid.

LEGS recognises the critical importance of global climate change and the urgent need to reduce greenhouse gas emissions. At the same time, it regards livestock-related support to poor and at-risk livestock keepers affected by humanitarian crises as entirely appropriate. The policy debates rely almost exclusively on data from large-scale commercialised livestock production systems, which are not the focus of LEGS (see *Principle 3: Climate change and the environment* in *Chapter 2*).

AMR

Antimicrobial resistance (AMR) is a key emerging issue. It is often associated with overuse and misuse of antibiotics as part of intensification of animal production, and with unregulated sales. When antibiotics are provided as part of emergency response packages, LEGS guidance recommends that every effort must be made to ensure they are used properly to reduce AMR.

One Health

The One Health approach brings together three topics: human health, animal health and environmental health. These topics are sometimes considered separately but are inextricably linked. Threats to these interconnected systems include human population growth, natural resource exploitation, large infrastructure development, and livestock production intensification and expansion. Risks also come from increased global movement of humans, animals (including wild animals) and animal products.

LEGS framework
Introduction to LEGS

Veterinary public health has a long history of considering links between people, animals and the environment. Many old and current zoonosis prevention approaches cover multiple aspects of disease transmission and risks. There is also now increasing acceptance of a One Health approach. In emergency response contexts, evidence of its positive impact on livelihoods is limited. However, LEGS guidance is that agencies must consider how to reduce the risk of public exposure to animal-borne diseases in emergency contexts.

Humanitarian-development-peace nexus

The trend towards complex emergencies, characterised by widespread and severe political instability and related conflict, is increasing. A relatively high proportion of humanitarian funding is now spent on complex emergencies. Adverse and more frequent weather-related hazards, like drought or severe storms, affect at-risk people more frequently than before and may compound these complex emergencies.

The Humanitarian-Development-Peace Nexus approach seeks to address complex emergencies by coordinating response efforts from the humanitarian, development and peace sectors. For LEGS, responding to complex humanitarian crises requires recognising the links between immediate and long-term needs, and the necessity of coordination (see Principle 5: Coordinated responses, in Chapter 2).

Livestock insurance

Index-based livestock insurance schemes are complementary to LEGS in that payouts aim to protect livestock assets through the provision of feed, water, shelter and so on. While there is a growing interest in such schemes, insurance policyholders make up only a very small proportion of total livestock producers in the countries where these schemes operate. The crisis-affected communities that LEGS targets fall largely outside insurance companies' priorities. There is not yet enough research on the impact of livestock insurance on livelihoods to enable LEGS to formulate guidance.

How does LEGS relate to Sphere and the Humanitarian Standards Partnership?

Established in 1997, Sphere developed global minimum standards for humanitarian response, with the aim of improving the quality and accountability of humanitarian action. The minimum standards are presented in the Sphere Handbook, with the fourth edition published in 2018. Sphere focuses on the core elements of humanitarian responses. It has a legal foundation and a strong rights-based framework. Sphere's foundations are **The Humanitarian Charter**, the **Protection Principles** and the **Core Humanitarian Standard** on Quality and Accountability.

The first edition of LEGS was published in 2009. Since then, it has been closely aligned with the humanitarian foundations of the Sphere Handbook, including its guidance on human protection. The use of LEGS means commitment to its foundations in Sphere, with important implications for response design and implementation. This commitment also affects the monitoring, impact evaluation and accountability of livestock-related interventions across all types of emergencies (see *Chapter 3: Emergency response planning*).

LEGS is one of the founder members of the Humanitarian Standards Partnership (HSP), a collaboration among standards initiatives that started in 2016. Sphere acts as the foundation for all the standards initiatives that comprise the HSP, which currently includes:

- The Sphere Handbook;
- Minimum Standards for Camp Management;
- Minimum Standards for Child Protection in Humanitarian Action;
- Humanitarian Inclusion Standards for older people and people with disabilities;
- Minimum Standards for Education;
- Livestock Emergency Guidelines and Standards;
- Minimum Economic Recovery Standards;
- Minimum Standard for Market Analysis;
- The Standards for supporting crop-related livelihoods in emergencies;
- CHS Alliance (HSP associate member and co-curator of the Core Humanitarian Standard).

LEGS framework
Introduction to LEGS

> **Box 1.3**
> # LEGS commitment to the HSP
>
> ### The Humanitarian Charter
>
> The Humanitarian Charter represents a commitment from agencies to follow humanitarian principles and be held accountable for their work. The Charter includes three bodies of international law: human rights law, refugee law and international humanitarian law. It also includes two fundamental moral principles, the principle of humanity and the principle of humanitarian imperative. The Humanitarian Charter proposes that humanitarian agencies recognise three rights: the right to life with dignity; the right to receive humanitarian assistance; and the right to protection and security.
>
> The LEGS commitment to The Humanitarian Charter is reflected in the human rights basis of LEGS, and in its advocating for humanitarian assistance and protection that ensures inclusion. LEGS recognises and responds to marginalised social groups within crisis-affected areas. These may include children (in particular, separated, unaccompanied or orphaned children), women, older people, people with disabilities, or groups marginalised because of religion, ethnic group, caste or gender identity. *(See also Principle 2: Community participation, in Chapter 2).*
>
> ### Protection Principles
>
> Agencies responding to emergencies are responsible for ensuring that their interventions do not increase risks to affected communities. To ensure this, they follow the four Protection Principles described in detail in the Sphere Handbook:
>
> - enhance people's safety, dignity and rights, and avoid exposing them to further harm;
> - ensure people's access to impartial assistance according to need and without discrimination;
> - assist people to recover from the physical and psychological effects of threatened or actual violence, coercion or deliberate deprivation;
> - help people to claim their rights.
>
> See Chapter 2 *Appendix 2.1* for details of how LEGS supports the Protection Principles.

Core Humanitarian Standard on Quality and Accountability

All the HSP initiatives seek to improve quality and accountability for emergency response. They do this by adhering to the **Core Humanitarian Standard** on Quality and Accountability (CHS) and its nine commitments. Each commitment has a quality criterion, key actions and organisational responsibilities.

Commitment to the CHS is at the organisational level. Therefore, the commitment applies across all types of humanitarian response, including the livestock-related responses that an organisation supports. Assessing the accountability of livestock-related responses therefore falls under generic CHS commitments and quality criteria. LEGS alignment with the CHS is shown through the implementation of the LEGS Principles (see *Chapter 2*).

The specific components of how LEGS Principles support the CHS commitments are summarised in *Appendix 2.2*.

What sources of information exist for understanding more about LEGS?

Practical advice for carrying out LEGS technical response interventions

LEGS aims for better quality and accountability in emergency livestock responses. It achieves this by providing standards and guidance to support good, evidence-based practice and decision-making. LEGS does not, however, give detailed information on how to implement livestock interventions, nor on national emergency response mechanisms, nor on how to access funding for emergency responses.

In 2016, the Food and Agriculture Organization of the United Nations (FAO) published a practical manual on how to carry out livestock interventions in emergencies. This manual is designed to complement LEGS:

FAO (2016) *Livestock-related interventions during emergencies – The how-to-do-it manual.* Edited by Philippe Ankers, Suzan Bishop, Simon Mack and Klaas Dietze. FAO Animal Production and Health Manual No. 18. Rome, https://www.fao.org/3/i5904e/i5904e.pdf

Each of the LEGS Handbook technical chapters *(Chapters 4, 5, 6, 7, 8, 9)* refers to the relevant practical content in the FAO manual. The further reading references at the end of each LEGS technical chapter also point to other sources of practical guidance.

Animal welfare

World Organisation for Animal Health (no date) *What we do: Animal Welfare.* See https://www.woah.org/en/what-we-do/animal-health-and-welfare/animal-welfare/

Recommendations for animal welfare are included in the WOAH *Terrestrial Animal Health Code* (see link above). Many national organisations produce and disseminate animal welfare standards and guidelines. For example: https://science.rspca.org.uk/sciencegroup/farmanimals/standards

Sustainable livelihoods

There are a large number of publications on sustainable livelihoods, including:

Chambers R. and Conway, G. (1991) *Sustainable rural livelihoods: practical concepts for the 21st century,* IDS Discussion Paper no. 296, https://www.ids.ac.uk/publications/sustainable-rural-livelihoods-practical-concepts-for-the-21st-century/

DFID (1999) *Sustainable livelihoods guidance sheets,* DFID 1999–2001, https://www.livelihoodscentre.org/documents/114097690/114438878/Sustainable+livelihoods+guidance+sheets.pdf/594e5ea6-99a9-2a4e-f288-cbb4ae4bea8b?t=1569512091877

Serrat O (2008) *The sustainable livelihoods approach,* ADB Knowledge Solutions paper, https://www.adb.org/sites/default/files/publication/27638/sustainable-livelihoods-approach.pdf

Levine S (2014) *How to study livelihoods: bringing a sustainable livelihoods framework to life,* Working Paper 22, Secure Livelihoods Consortium, https://securelivelihoods.org/wp-content/uploads/How-to-study-livelihoods-Bringing-a-sustainable-livelihoods-framework-to-life.pdf

Pandemics

The following publications give guidance on livestock programming in the pandemic context:

FAO (2020) *Guidelines to mitigate the impact of the COVID-19 pandemic on livestock production and animal health*, https://www.fao.org/in-action/kore/publications/publications-details/en/c/1277631/

FAO (2020) *Guidance note: risk communication and community engagement. Coronavirus disease 2019 (COVID-19) pandemic*, https://www.fao.org/policy-support/tools-and-publications/resources-details/en/c/1306987/

Global Food Security Cluster (2020) *Guidance for emergency livestock actions in the context of COVID-19: addressing emerging needs related to the pandemic and reprogramming on-going critical activities*, https://fscluster.org/coronavirus/document/guidance-emergency-livestock-actions

Catley, A (2020) *COVID-19, livestock and livelihoods: a discussion paper for the Livestock Emergency Guidelines and Standards* (LEGS), https://www.livestock-emergency.net/wp-content/uploads/2021/01/LEGS-COVID-19-Discussion-Paper-single-pages.pdf

LEGS (2020) *COVID-19 response: LEGS guidance note, (LEGS)*, https://www.livestock-emergency.net/wp-content/uploads/2020/04/LEGS-COVID-19-Guidance-Note-28-April-2020.pdf

Veterinary medicines and antimicrobial resistance

Drawing on the experiences and conclusions of the LEGS Operational Research Project, this discussion paper looks at supply chain issues and the global problem of AMR:

Hufnagel, H (2020) *The quality of veterinary pharmaceuticals: a discussion paper for the Livestock Emergency Guidelines and Standards* (LEGS), https://www.livestock-emergency.net/wp-content/uploads/2020/11/LEGS-Discussion-Paper-The-Quality-of-Veterinary-Pharmaceuticals.pdf

Hufnagel, H (2022) *The quality of veterinary pharmaceuticals:* LEGS technical brief, LEGS, https://www.livestock-emergency.net/wp-content/uploads/2022/06/Quality-Veterinary-Pharma-Brief_June-2022.pdf

LEGS framework
Introduction to LEGS

Humanitarian Standards Partnership

The Humanitarian Standards Partnership currently comprises Sphere and nine other initiatives.

The Sphere Handbook, Sphere Association, https://spherestandards.org/handbook-2018/

Minimum Standards for Child Protection in Humanitarian Action (CPMS), The Alliance for Child Protection in Humanitarian Action, https://alliancecpha.org/en/CPMS_home

Livestock Emergency Guidelines and Standards (LEGS), https://www.livestock-emergency.net/download-legs/

Minimum Economic Recovery Standards (MERS), SEEP Network, https://seepnetwork.org/MERS

Minimum Standards for Education, Inter-Agency Network for Education in Emergencies (INEE), https://inee.org/resources/inee-minimum-standards

Minimum Standard for Market Analysis (MISMA), CALP Network, https://www.calpnetwork.org/publication/minimum-standard-for-market-analysis-misma/

Humanitarian inclusion standards for older people and people with disabilities, Age and Disability Capacity Programme (ADCAP), https://www.helpage.org/what-we-do/emergencies/adcap-age-and-disability-capacity-building-programme/

Minimum Standards for Camp Management Global Camp Coordination and Camp Management (CCCM) Cluster, https://cccmcluster.org/resources/minimum-standards-camp-management

Standards for Supporting Crop-related Livelihoods in Emergencies (SEADS), https://seads-standards.org/

Core Humanitarian Standard (CHS), CHS Alliance (HSP Associate member and co-curator of the CHS), https://corehumanitarianstandard.org/

What further information is available on topics that LEGS does not cover?

Animal disease emergency guides

LEGS does not cover responses to animal disease epidemics, even though these are sometimes very disruptive to livestock-based livelihoods. There is detailed advice available on this from other sources. LEGS also does not cover transboundary animal disease control. Other internationally accepted guidelines cover this specifically.

The World Organisation for Animal Health (founded as OIE) (2021) *Terrestrial Animal Health Code*, https://www.woah.org/en/what-we-do/standards/codes-and-manuals/terrestrial-code-online-access/

Guides from the FAO Emergency Prevention System for Animal Health (EMPRES-i), https://empres-i.apps.fao.org/

The Food and Agriculture Organization of the United Nations (FAO) produces disease-specific guides, such as:

Tuppurainen, E., Alexandrov, T. and Beltrán-Alcrudo, D. (2017) 'Lumpy Skin Disease: A Field Manual for Veterinarians', *FAO Animal Production and Health Manual No. 20*, FAO, Rome, https://www.fao.org/3/i7330e/i7330e.pdf

FAO and OIE/WOAH (2020) *Global control of African swine fever: a GF-TADs initiative*, 2020–2025. Paris, https://www.woah.org/app/uploads/2021/06/global-control-of-african-swine-fever-a-gf-tads-initiative-2020-2025.pdf

Aquaculture and apiculture

LEGS does not cover aquaculture or beekeeping (apiculture), but FAO provides the following guidance:

Cattermoul, B., Brown, D. and Poulain, F. (eds) (2014) *Fisheries and aquaculture emergency response guidance*, FAO, Rome, https://www.fao.org/3/i3432e/i3432e.pdf

FAO, IZSLT, Apimondia and CAAS (2021) *Good beekeeping practices for sustainable apiculture,* FAO Animal Production and Health Guidelines No. 25, Rome, https://www.fao.org/documents/card/en/c/cb5353en

LEGS framework
Introduction to LEGS

Companion animals

Given its humanitarian and livelihoods perspectives, LEGS does not explicitly cover companion animals. However, many of the LEGS standards and guidance notes apply to companion animals. Links to further resources that include companion animals in emergencies are available at: https://www.nal.usda.gov/legacy/awic/disaster-planning

Livestock insurance

Aklilu, Y (2020) *Livestock insurance: a discussion paper for the Livestock Emergency Guidelines and Standards*, LEGS, https://www.livestock-emergency.net/wp-content/uploads/2021/01/LEGS-Insurance-Discussion-Paper-single-page.pdf

See also case studies for livelihoods-based emergency interventions at: https://www.livestock-emergency.net/resources/case-studies/

LEGS framework
Introduction to LEGS

LEGS framework
LEGS Principles

Chapter 2: LEGS Principles

48	**Introduction**
48	**Principle 1: Supporting livelihoods-based programming**
52	**Principle 2: Ensuring community participation**
55	**Principle 3: Responding to climate change and protecting the environment**
57	**Principle 4: Supporting preparedness and early action**
59	**Principle 5: Ensuring coordinated responses**
61	**Principle 6: Supporting gender-sensitive programming**
63	**Principle 7: Supporting local ownership**
66	**Principle 8: Committing to monitoring, evaluation, accountability and learning (MEAL)**
69	**Appendix 2.1: LEGS alignment with the Sphere Protection Principles**
70	**Appendix 2.2: How LEGS Principles support the Core Humanitarian Standard Commitments**
73	**References and further reading**

Photo © Sippakorn Yamkasikorn/Unsplash

LEGS framework
LEGS Principles

Chapter 2: LEGS Principles

Introduction

This chapter describes the eight LEGS Principles. The LEGS Principles are evidence-based or are derived from many years' experience of implementation, and related evaluations. They also take account of reviews and practitioner experience. The LEGS Principles are complementary to the Sphere foundations. They cover fundamental issues and ways of working that are the basis for effective livestock-related responses in humanitarian crises (see *Figure 2.1*). Appendix 2.1 and 2.2 show how the LEGS Principles relate to the Sphere Protection Principles and the Core Humanitarian Standard (CHS).

The LEGS Principles apply across all humanitarian contexts and all LEGS technical interventions; however, the practical application of each principle will vary according to context. Humanitarian assistance involves managing dilemmas, and although an agency might understand and commit to a principle, specific political, security or other contextual issues may affect the practical application of the principle. Implementing agencies should apply all of the LEGS Principles during emergency interventions. They should also apply them to their ways of working and their internal structures.

Principle 1: Supporting livelihoods-based programming

The benefits of livelihoods-based livestock programming

The LEGS Principle of supporting livelihoods-based programming aligns with the general humanitarian aim of 'saving lives and livelihoods'. It recognises that humanitarian response should include efforts to protect people's livelihoods and support post-emergency recovery. The livelihoods-based programming recommended by LEGS is also consistent with the rights-based approach in the foundation chapters of Sphere. In communities where livestock produce food and income, livestock keepers have the right to emergency support to protect and rebuild their livestock.

Figure 2.1: Sphere foundations, LEGS Principles and the LEGS technical interventions

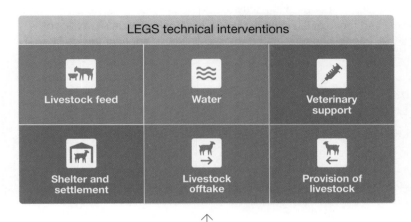

Principles support the LEGS technical interventions

Foundations underpin the LEGS Principles

LEGS framework
LEGS Principles

For livestock keepers facing an emergency, the main ways in which livestock contribute to livelihoods are summarised in *Box 2.1*. The two primary determinants of people's capacity to survive emergencies are their financial capital and their social networks. Here, livestock provide key types of financial capital and play important roles in social connectedness in rural communities.

Box 2.1
Livestock and livelihood assets in emergencies

Livestock and financial capital

In many lower- and middle-income countries, livestock are both a source of cash income and a form of financial saving in rural communities. Income from the sale of livestock or livestock products can be a substantial proportion of total household income. Similarly, livestock can be the most important type of financial savings. They provide both savings growth (as livestock reproduce) and a flexible disposable asset to meet household income needs, including at times of stress. Working animals also provide major economic benefits, such as enabling the ploughing of land or transporting goods to markets.

Livestock and social capital

During emergencies people's survival can depend on their capacity to draw on assistance from relatives and friends. In rural communities, local social support systems often rely heavily on livestock-based transactions, such as loans or gifts of livestock or livestock products. These systems apply during normal periods, during emergencies and as part of recovery after emergencies. Pastoralist and agro-pastoralist communities have elaborate systems for restocking households that have lost animals due to drought, disease epidemics or other events.

Livestock and human capital

Livelihood frameworks typically refer to human capital as a household's health status, level of education and capacity to work. Animal-source foods such as milk and eggs have very high nutritional value and are particularly important for young children, and for pregnant and breastfeeding women. In children, these foods contribute to both physical and cognitive growth. The direct consumption of animal-source foods from livestock owned by poorer households may be the only way for them to access or afford such food.

Participatory approaches should be used to identify options for supporting livestock during emergencies (see *Chapter 3: Emergency response planning*). When they are, livestock keepers typically prioritise those LEGS interventions that lead to one or more of the livelihood contributions shown in *Box 2.1*. They especially focus on the protection of core livestock assets. These livelihood contributions are also reflected in the LEGS livelihoods objectives.

LEGS technical interventions support crisis-affected communities to:

1. **obtain immediate benefits using existing livestock assets;** and/or
2. **protect key livestock assets;** and/or
3. **rebuild key livestock assets.**

Livelihoods-based programming and local services and markets

The livelihoods-based approach recommended by LEGS protects livestock as an essential livelihood asset. It also aims to support the services, markets and systems that livestock keepers need as they recover from emergencies. This approach takes account of lessons learned from negative experiences relating to the free distribution of livestock medicines, feed and other inputs by NGOs and governments during emergencies. Free inputs can undermine local private sector actors, disrupt local market systems, or introduce inappropriate products. Rather than working in isolation away from private veterinary workers and veterinary paraprofessionals, feed suppliers and traders, LEGS recommends the inclusion of these operators. This way they are involved in the design and implementation of LEGS technical interventions. More detailed guidance is provided in the LEGS technical chapters.

Livelihoods-based programming and humanitarian versus development support

It is important, when working with the private sector or governments on LEGS technical interventions, to distinguish between types of support. These include support that is essential during an emergency or in the immediate post-emergency phase to protect or rebuild livestock assets. Other types of support involve long-term development activities. This distinction is needed to avoid the diversion of humanitarian resources towards development work. It is consistent with CHS Commitment 9 on the effective, efficient and ethical use of resources (see *Chapter 1: Introduction to LEGS* and *Appendix 2.2*, which shows how LEGS Principles link to the CHS).

LEGS framework
LEGS Principles

From a humanitarian perspective, this distinction often means placing most emphasis on the basic needs of livestock, such as feed, water and clinical veterinary care. This includes related input suppliers and service providers. Support to systems such as disease surveillance, and consideration of long-term and complex policy and institutional constraints to livestock development, are lower priorities. These should be addressed during 'normal', non-emergency periods. The use of participatory needs assessment and response design with communities and local service providers is a practical way to identify and prioritise types of livestock-related humanitarian support as opposed to developmental needs.

Making clear distinctions between humanitarian and development priorities is not always straightforward. In complex emergencies, humanitarian programmes can become long-term. Over time, livestock professionals will typically request support for disease surveillance, veterinary public health or other public sector activities such as policy or regulatory reform. The extent to which using humanitarian resources for these activities is justified depends on the context (including whether the basic needs of livestock, such as feed, water and clinical veterinary care, are already being met).

Principle 2: Ensuring community participation

Community participation in emergency livestock projects

Community participation is a fundamental aspect of effective and ethical humanitarian assistance. It relates directly to the first principle of the Humanitarian Charter, and the commitments and quality criteria of the CHS (see *Appendix 2.2*). Community participation is a requirement to meet the Sphere Protection Principles. It is also needed, for example, to ensure that interventions do no harm and do not expose people to risk of violence or abuse (see *Box 1.3* in *Chapter 1: Introduction to LEGS*).

In emergency livestock interventions, there is a long history of agencies working closely with communities to design and implement responses. There is also a body of evidence showing the links between the quality of community participation and livelihood impacts. Participation includes respect for indigenous knowledge on the local environment and grazing management, livestock husbandry and diseases, and customary social systems and networks that depend on livestock transactions. This knowledge has substantial practical value when identifying appropriate emergency interventions. It is particularly relevant in communities where livestock are central to household economies.

Incorporating indigenous knowledge on livestock production and disease is also critical where statistics and data on these subjects are not available or are out of date. Limited data on livestock matters is common in more remote and marginalised rural communities that depend heavily on livestock. Local customary institutions and leaders can play important roles in mobilising and organising livestock keepers for dialogue and meetings where needs are identified and prioritised, and where appropriate interventions designed.

Ensuring active community participation

There is widespread use of participatory approaches by agencies involved in emergency livestock-related responses. Despite this, there can be mixed understanding of what 'community participation' means, and the types of participation that are effective in humanitarian contexts. LEGS recommends that agencies aim to achieve **active participation** (see Table 2.1), because this has been shown to be achievable in emergency livestock-related contexts. It is also associated with strong livelihood benefits. Livestock keepers should be included in all stages of the project cycle: assessment, design, planning, implementation and also monitoring, evaluation, accountability and learning (MEAL).

As part of active community participation, it is important to consider different social, ethnic and wealth groups within communities. The use and ownership of livestock will often vary according to wealth, gender, age or other factors. Effective identification, design and implementation of livestock interventions therefore requires the specific involvement of marginalised or at-risk groups who keep livestock. It should also include marginalised or at-risk groups who benefit from access to livestock or livestock products. Such groups may include female-headed households, internally displaced people (IDP), refugees, or stateless communities.

The initial assessment should aim to understand how interventions might be targeted at different groups with different potential impacts. Agencies need to be sensitive to these differences and aware that men or wealthier people might dominate local leadership and customary institutions. The initial assessment should also consider how interventions will impact livestock keepers of different age groups. This is because their roles and responsibilities may vary as their capacity for management and income generation changes with age.

LEGS framework
LEGS Principles

It is important to address barriers to the participation of women, older people and at-risk groups (including men and women with disabilities) at both the assessment and implementation stages – see Humanitarian inclusion standards for older people and people with disabilities (HIS). Agencies should also consider potential impacts on children and adolescents, including both child labour and child protection risks in insecure environments – see Child Protection Minimum Standards (CPMS) Principle 3, *Children's participation*, and Standard 22, *Livelihoods and child protection*.

Active participation requires close proximity to communities, and good communication between agency staff and local people. One effect of Covid-19 restrictions is that, in many countries, direct interaction with communities has become more challenging. Experiences of operating during Covid-19 restrictions are still emerging. Yet a general lesson is that local organisations and networks are better able to adapt to changing Covid-19 contexts than international agencies *(see also LEGS Principle 7: Supporting local ownership)*.

Table 2.1: Types of community participation in emergency livestock-related interventions

Type of community participation	Relevance to LEGS
Passive participation: Communities participate by being told what has been decided or has already happened. Involves unilateral announcements by an external agent or management without listening to people's views. The information belongs only to external professionals.	Not consistent with Sphere foundations, and not recommended by LEGS.
Participation by consultation: Communities participate by being consulted or by answering questions. External agents define the problems and the information-gathering processes and so control analysis. The consultative process does not concede any share in decision-making, and professionals have no obligation to take on board people's views.	As above.
Participation for material incentives: Communities participate by contributing resources such as labour in return for material incentives (e.g. food, cash).	As above.

continued over

LEGS framework
LEGS Principles

Type of community participation	Relevance to LEGS
Functional participation: External agencies view community participation as a means to achieve intervention goals. People participate by forming groups to meet predetermined project objectives. They may be involved in decision-making, but only after external agents have already made major decisions.	As above.
Active participation: People participate in joint analysis, in the development of action plans and the formation or strengthening of local institutions. Participation is seen as a right rather than a means to achieve project goals. Since groups take control of local decisions and determine how available resources are used, they have a stake in maintaining structures or practices.	**This aligns with Sphere foundations; it reflects good practice in emergency livestock projects; LEGS recommends it.**
Self-mobilisation: People participate by taking initiatives independently of external actors. They develop contacts with external actors for the resources and technical advice they need, but retain control over how resources are used. Self-mobilisation can spread if governments and NGOs provide an enabling framework of support.	There is very limited documentation on self-mobilisation in emergency livestock initiatives.

Principle 3: Responding to climate change and protecting the environment

Global debates on livestock and climate change

LEGS acknowledges the importance of global debates on the climate change impacts of livestock production. However, it also takes account of the substantial livelihood benefits from livestock for the hundreds of millions of people currently living in areas affected by emergencies. In the medium to long term, there is a general lack of alternative and feasible livelihood options for rural communities who depend heavily on livestock. This dependence includes the direct consumption of livestock products. Livestock provide vast nutritional benefits for populations with high levels of malnutrition, and where foods rich in protein and other essential nutrients are not affordable.

LEGS also recognises that the research currently underpinning debates on livestock systems and human diets relies almost exclusively on data from

LEGS framework
LEGS Principles

large-scale commercialised livestock production systems. These are fundamentally different from the small scale or extensive systems used by poorer communities in low-income countries. Policymakers need to consider diverse livestock and livelihood systems and contexts. They must avoid broad and simplistic solutions that are mainly relevant to commercial livestock systems in industrialised countries.

Climate change impacts on livelihoods

Climate change has major direct and indirect impacts on livestock keepers, their livestock and their livelihoods in a range of interrelated ways. The gradual changes in climate include temperature and rainfall changes that affect animal performance. They also involve changes in water availability, patterns of animal disease, and the species composition of rangelands – for example, the impact of invasive species such as *Prosopis juniflora* in India and East Africa. Indirect impacts include changes in the price and market availability of both animal feed and human food. People might also change the ways they use land, cultivating biofuels, for example, as a response to climate change. Many small-scale livestock keepers use mixed farming systems, combining livestock and crop production on small, and often declining, areas of land. Crop systems are also affected by climate change, meaning that these producers have to manage changes to both livestock and crop systems (see SEADS).

In general, poorer livestock producers are less able to adapt to climate-related pressures. This is especially so if this adaptation requires higher input costs, or financial investment or credit. Many smaller producers are becoming increasingly at risk, so when natural hazards occur, the impacts are more severe. Changes in the frequency and severity of extreme weather events mean that poorer producers face not only the effects of climate change but also increasing exposure to slow-onset or rapid-onset climate-related emergencies. These issues become compounded in complex emergencies, when superimposed on chronic political instability and conflict.

The increasing impacts of climatic trends on small-scale or at-risk livestock keepers in lower-income and middle-income countries make effective disaster early warning systems and early response more and more important – see LEGS *Principle 4: Preparedness*.

Livestock keeping and local environmental management

Most small-scale producers and livestock keepers use low-input, low-output livestock production systems. These depend heavily on naturally available pasture, browse and water, and to a lesser extent on crop residues.

Therefore, environmental management is central to livestock production and the related financial, social and nutritional benefits. However, the long-term policy and development issues around environmental management are complex and beyond the scope of LEGS. For poorer livestock keepers, a key constraint is often uncertain land tenure. This comes with the associated risks of investment in land that can easily be appropriated for commercial or political interests.

To support local environmental management, LEGS focuses on measures that are practical and achievable. It does this during emergencies and in the immediate post-emergency recovery phase. It is important to understand how emergencies can increase the risk of negative environmental impacts from livestock. For example:

- Reduced pasture, fodder and water due to drought can result in concentrations of livestock around declining water resources and localised overgrazing.
- Displaced people may move to camps with their livestock, resulting in unusually high livestock populations in confined areas. Although the provision of feed and water may sustain livestock in these situations, sanitary issues must be considered. Use of nearby grazing and water points already in use by local residents can lead to overuse and environmental damage.
- Displacement and restrictions on migration, because of conflict or other factors, limit the normal movement of animals and concentrate livestock. This may result in localised overgrazing and deterioration of animal health.

Further environmental considerations in some emergencies include the management of waste from livestock (particularly in camp settings). This could also involve the disposal of livestock waste following slaughter, and the disposal of livestock carcasses. Some emergencies, particularly those caused by flooding, can result in the deaths of tens of thousands of animals. This presents a considerable challenge if negative environmental and human health impacts are to be avoided.

During emergencies, veterinary interventions require environmental assessment. This includes assessing the careful use of certain drugs to control ectoparasites, especially in the form of dips and sprays, and the use of specific anthelmintic and anti-inflammatory medicines.

LEGS framework
LEGS Principles

Principle 4: Supporting preparedness and early action

LEGS evidence reviews and practitioner experience show clear links between early livestock-related response, livelihood benefits and cost-effectiveness. Most of the evidence on the value of early response focuses on drought (a slow-onset emergency). Despite this, the principle of preparedness and early action applies widely across most types of both slow-onset and rapid-onset emergencies. Many of the LEGS technical chapters include a standard on preparedness, which provides more details of activities that can support the application of this principle.

Where possible, the key elements of disaster risk reduction (DRR) should be supported in areas that are prone to emergencies. These elements include preparedness, contingency planning and early response. In complex emergencies, DRR will apply to the slow-onset droughts or rapid-onset events that may occur alongside the other factors causing the crisis. As described under LEGS *Principle 3: Responding to climate change and protecting the environment*, increasing climate variability means LEGS interventions need to strengthen preparedness and timely response.

In some regions, there is increasing use of flexible funding mechanisms in development projects. These enable rapid diversion of development funds towards emergency response. Sometimes called crisis modifiers, flexible funding has been shown to be particularly useful for supporting early livestock response to drought, as part of drought cycle management. When these responses are designed to meet LEGS objectives and involve local service providers or market actors, they fit well with LEGS *Principle 1: Livelihoods-based programming*.

A critical aspect of DRR and flexible funding is the integration of these approaches into long-term development programmes. Many effective emergency livestock responses have been implemented by aid agencies with long-term development experience in a particular area, based on contingency planning that is part of a development programme. Such plans and triggers for action are guided by knowledge of past humanitarian crises and the types of response that can be implemented within a given operational and funding context. However, these approaches also mean that the programme managers need in-depth experience of both development and humanitarian concepts – including early warning. They also need to show leadership and commitment to preparedness and early action, including incorporating flexible funding at the design stage, with donor support.

Active community involvement in developing early warning systems, contingency planning and flexible budgets is recommended for all types of emergencies. This is in line with LEGS *Principle 2: Ensuring community participation*. Preparedness planning should build capacity in local organisations and community institutions and enable local knowledge and priorities to guide early warning planning. In drought-affected areas, specific capacity strengthening in drought cycle management can help to ensure appropriate sequencing of interventions. Across all areas affected by climate-related emergencies, capacity for interpreting climate forecasts and using the information is important. Similar support is needed for local actors in non-climatic events such as earthquakes or volcanic eruptions.

Early warning, preparedness and early action are often viewed by implementing agencies as mainly technical activities. Yet effective early action depends heavily on the involvement of senior administrative staff. Such staff can advise on approval, procurement, recruitment or other procedures. They can also make recommendations about how to speed up, waive or otherwise apply these procedures to support rapid action. For example, pre-positioning of private sector operators such as transport companies, feed suppliers, or veterinary workers may be possible, with draft contracts prepared as part of the planning process.

A further aspect of preparedness is ensuring that implementing agency staff have adequate knowledge, experience and communications skills to deliver a quality emergency response. This includes technical competencies such as livestock management or veterinary training, as well as experience in participatory approaches and community engagement *(see Principle 2: Community participation*, and *Principle 7: Local ownership)*.

Principle 5: Ensuring coordinated responses

In most humanitarian crises, different interventions take place simultaneously. So coordination is vital to maximise impact, avoid duplication and ensure efficient use of resources. Coordination is usually led by government bodies or UN agencies. It can ensure that multiple agencies are mobilised, that they are committed to meeting common objectives and that they use humanitarian standards, such as LEGS.

LEGS framework
LEGS Principles

Experience has shown that coordination requires all partners to make a commitment of time and staff. Under the main coordinating effort, the creation of working groups for particular regions or for particular types of emergency may help to harmonise approaches. As well as agreeing roles and responsibilities, the partners can create linkages with livelihoods and ongoing development initiatives. Donors may also be well placed to encourage or even demand harmonisation of approaches by implementing agencies.

Two main types of coordination are most relevant to LEGS: coordination between livestock interventions, and coordination between livestock interventions and other sectors.

Coordination between livestock interventions

Given the range of emergency livestock interventions and the need to tailor them to specific subpopulations or at-risk groups, coordinated responses are critical. If different implementing agencies are providing different types of support, coordination is needed to avoid duplication and to ensure that an important type of support is not overlooked. For example, if a combined feed–water–livestock health response is needed, failure to provide one type of support risks the effectiveness of the other responses.

When different implementing agencies provide similar support, coordination should ensure approaches are harmonised and programming is consistent. For example, if agencies covering adjacent areas set different purchase prices for livestock offtake, or employ different distribution policies for the provision of livestock (free, loan, subsidised, and so on), the initiatives may undermine each other. In veterinary support, differing policies on cost recovery can weaken interventions and cause confusion among affected communities. In slow-onset emergencies such as drought, another aspect of the coordination effort should be to promote appropriate sequencing of interventions according to the stage of the drought.

Strong coordination can also enable agencies with limited expertise in livestock to receive technical guidance and support from more specialised agencies.

Coordination between livestock interventions and other sectors

Coordination between livestock interventions and other sectors requires attention to two main issues.

The need to time and prioritise livestock support in relation to critical humanitarian life-saving activities

In an emergency, the most urgent need may be to provide life-saving assistance to affected human populations. Such assistance should not be compromised or adversely affected by providing livestock assistance. In practice, this means that when emergency transportation, communication or other resources are limited, livestock teams and inputs should follow the food, shelter, water and health inputs required to assist people in need. For example, water delivery programmes should ensure that people's water needs are catered for before, or at the same time as, those of livestock, or make use of different quality water for the two groups.

Technical and logistical coordination with other sectors

Where possible, LEGS recommends that livestock interventions should be integrated with other types of humanitarian assistance to maximise impact and ensure efficient use of shared resources. Many farming systems in countries impacted by humanitarian crises are mixed systems, involving crop and livestock production. Standards for Supporting Crop-related Livelihoods in Emergencies are described in SEADS. There are various opportunities for coordinating livestock and crop activities, including joint assessments, harmonised timing of inputs and joint monitoring and evaluation.

In some livestock-keeping communities, children are involved in livestock care. This may impact their access to schooling. The INEE Minimum Standards for Education, which relate to education in emergencies, and the Minimum Standards for Child Protection (CPMS) provide important guidance, including ensuring flexible school calendars and a range of non-formal education options.

Even in situations where technical coordination with other sectors might not be relevant, there can still be opportunities for implementing agencies to share resources and coordinate from a logistical perspective. For example, trucks delivering aid supplies could be backloaded with livestock as part of livestock offtake; refrigerators might store both human and animal medicines; discarded or damaged items for human shelter could be used for animal shelter.

LEGS framework
LEGS Principles

Principle 6: Supporting gender-sensitive programming

Gender-sensitive programming is important in any emergency. This is because women and girls, men and boys, and people with other gender identities have access to different resources and different coping strategies. These resources and strategies need to be understood and recognised. Effective gender-sensitive programming in emergencies requires recognising that:

- The crisis often disproportionately increases women's, girls' and boys' labour burden, while simultaneously reducing their access to food, key assets and essential services.
- The risk of gender-based violence, exploitation and abuse is also heightened. This can influence women's, girls' and boys' ability to survive and recover from conflicts and humanitarian crises.

In emergency livestock responses, there are now examples of more inclusive approaches involving gender-sensitive programming, and related livelihoods or nutritional benefits. As a minimum, the role of livestock within women's livelihoods should be addressed as part of any emergency intervention. This is not least because children's health and nutrition are often closely associated with women's bargaining power, their control over livestock resources, and their control over their time and workload.

Gender-sensitive programming and LEGS technical interventions

Gender-sensitive programming not only focuses on women, but on an equitable focus of support for men, women and children. However, within LEGS interventions there are some key issues for women specifically. These include protecting women's livelihood assets during emergencies, identifying their priorities in livestock management, and recognising the cultural and economic barriers that often impact their livelihoods.

Evidence suggests that for emergency livestock responses, a sound gender-sensitive understanding of roles, rights and responsibilities relating to livestock is critical. This includes issues of ownership and control of livestock as key assets. In many livestock-keeping societies, control over livestock is more than a simple concept of 'ownership'. For example, women or children may be responsible for young stock but not for adult stock. Or women may control livestock products such as milk, butter, hides and skins as part of their overall control of the food supply. Men, meanwhile, may have rights over the sale, bartering or gifting of the animal itself. Following the

LEGS technical interventions requires consideration of women's, men's and children's daily and seasonal roles and responsibilities. This includes their access to and control of the various livestock species and age categories. For example, interventions may focus on large livestock species, ignoring the important role that small species, such as poultry, play in women's livelihoods.

It is also important to be aware that women and girls are often marginalised and face restrictions on their mobility in many livestock-keeping societies. This reduces their access to livelihood inputs, livestock resources and basic services. Other barriers include their uneven family and social responsibilities, unequal access to protective services and legal mechanisms, and inadequate political power at local or national levels.

In contrast, during some emergency contexts, women take up more powerful positions within livestock management – for example, if men have migrated to find work. It is important to ensure LEGS technical interventions do not undermine new roles, but instead strengthen them according to context. There can be unique opportunities created by an emergency for challenging conventional gender roles. These are best addressed by working with relevant (sometimes informal) community-based groups or leaders to address women's subordinate roles.

The initial assessment *(Chapter 3: Emergency response planning)* covers participatory dialogue with men and boys, and women and girls, to understand their preferences for livestock-related support. It also involves the collection of gender-disaggregated information, including about the impact of the emergency on different genders. The potential impact of any livestock intervention on men's, women's and children's workloads and on their control of livestock resources needs to be clearly understood. This will improve the design, implementation and impact of the intervention. Changing gender roles, responsibilities and status also need to be taken into account. Implementing agencies should recognise that participatory dialogue with women is often easier if the agency staff are themselves women. The time and resources available for planning, implementation and monitoring of emergency interventions are limited. Yet gender analyses are relatively easy to apply and can result in significant improvement to the approach and impact.

LEGS framework
LEGS Principles

Principle 7: Supporting local ownership

Supporting local ownership: experiences from livestock interventions in emergencies

The LEGS Principle of supporting local ownership draws on decades of experience of emergency livestock interventions. These interventions have involved joint technical decision-making between external agencies and local actors, and the development of equal technical working relationships with local government and others. There are also many examples of supporting local ownership. These include supporting governments to coordinate, monitor and evaluate activities; working with communities to design, implement and evaluate interventions; and supporting governments to develop minimum standards, design and run training courses, and revise policies. These activities have recognised and respected the roles of government and local NGOs, while aiming to strengthen capacities to prepare for and respond to future emergencies.

These livestock interventions align closely with the international humanitarian community's adoption of the 'localisation' concept. For example, the Organisation for Economic Co-operation and Development (OECD) describes the process of localising humanitarian response as 'recognising, respecting and strengthening the leadership by local authorities and the capacity of local civil society in humanitarian action, in order to better address the needs of affected populations and to prepare national actors for future humanitarian responses'. LEGS recognises that communities, local organisations and authorities are often the first to act in a crisis. They also have in-depth knowledge of the situation and specific needs.

Some emergency livestock initiatives have developed relationships with local partners so that technical decision-making has been transferred to them. An important challenge for the concept of supporting local ownership is the limited extent to which local organisations have taken control of budgets. Nor have they designed and implemented interventions independently or with full 'ownership' and control of resources. This situation relates to the policies of aid donors and international NGOs and cuts across the humanitarian sector. The sector-wide nature of the challenge indicates that livestock interventions alone cannot change how the wider humanitarian sector operates.

LEGS approach for supporting local ownership

Taking account of the issues and experiences outlined above, the LEGS Principle for supporting local ownership has the following main elements:

- Emergency livestock interventions should continue to build on past experience of developing equal technical partnerships with local actors. They should shift analysis and decision-making towards communities, and to local government and civil society.

- Supporting local ownership is aligned with *Principle 2: Ensuring community participation*, and LEGS tools encourage a locally designed (and ideally locally controlled) response. LEGS also recognises that options for supporting more local ownership are context-specific. In some contexts, a more traditional approach, dominated by international actors, is the only option for providing livestock-related support at scale.

- For all its technical interventions, LEGS promotes the delivery of responses under the leadership of national and local authorities. It also encourages responses led by community-based organisations *(see References and further reading)*. Additionally, LEGS supports enabling local authorities to coordinate livestock interventions across agencies; and developing minimum standards and guidelines that are adapted to local contexts.

- Emergency livestock interventions provide an opportunity to demonstrate how local NGOs and local government can take more ownership and control (for example by defining local financial needs, and receiving, and accounting for, external funding). Documenting these cases will provide further examples of local ownership in practice, and of its impacts and efficiencies. This will enable emergency livestock practitioners and organisations to lobby for wider changes to the humanitarian system from a position of practical experience and strengthened by examples of positive cases.

LEGS framework
LEGS Principles

Principle 8: Committing to monitoring, evaluation, accountability and learning (MEAL)

The benefits of implementing MEAL as part of emergency livestock response

The final LEGS Principle is the need to establish a monitoring, evaluation, accountability and learning (MEAL) system and commit to its implementation. A well-planned and executed MEAL system ensures emergency livestock responses are effective. It also helps to achieve the LEGS livelihoods objectives, and supports learning and continuous improvement. Effective MEAL promotes greater accountability for affected communities and is also a commitment under the Humanitarian Standards Partnership (HSP).

Monitoring provides the key information needed to assess progress when implementing an emergency livestock response. It is useful for checking whether the initial assessment is still valid or if changes in the operating context need to be considered. Critically, effective monitoring enables real-time adjustments to implementation. Evaluations usually assess whether project objectives have been achieved, and if not, why not. Impact evaluations or impact assessments look specifically at the impacts of responses. In the case of livestock responses, they measure key livelihood indicators such as livestock mortality, or income and food derived from livestock. When objectives include a clear reference to expected livelihood impacts, evaluations inherently include measures of impacts. Impact evaluations are especially useful for understanding livelihood impacts, and to identify positive and negative lessons. This contributes to improvements in the design and impact of future responses.

Being accountable and establishing a system for sharing learning require a commitment of time and effort. Gathering relevant information and feedback, analysing it and sharing lessons learned with people affected by a crisis promote accountability. Sharing learning helps improve future programming across the livestock and humanitarian sectors as a whole. LEGS adheres to Commitment 7 of the CHS, which requires humanitarian actors to continuously learn and improve. Whether an implementing agency is a specialised livestock organisation or multi-sectoral, it should commit to the CHS and have a system in place to ensure accountability against the CHS Commitments.

Important considerations for implementing MEAL in technical interventions

Many organisations may already have a MEAL system in place. Important considerations for establishing a MEAL system for livestock emergency response are:

- **Investment:** For effective MEAL, investment in staff skills and time is needed. Staff require sufficient competence in participatory techniques to facilitate high-quality information collection as well as appropriate understanding of livestock-based livelihoods. Given the urgency of most emergency response contexts, any required capacity strengthening or exchange is best made ahead of time, as part of preparedness planning.
- **Monitoring:** While recognising the challenges in operating in emergency contexts, monitoring systems should be in place as soon as LEGS interventions begin. Monitoring needs to be frequent enough to enable rapid detection of required changes and modifications that need to be made to implementation.
- **Prioritising impact evaluation:** Implementing agencies should allocate time and funding for impact evaluation towards the end of, or soon after, a response. Research shows that participatory impact evaluation produces good evidence of the livelihood impacts of emergency livestock responses. Such evaluation involves communities directly in assessment, learning and accountability.
- **Learning mechanisms:** For learning to take place, data collection alone is not sufficient. Mechanisms need to be established to help analysis and sharing of the monitoring and evaluation information.

Alignment with the other LEGS Principles

LEGS commitment to effective MEAL aligns with and supports the other LEGS Principles, in particular:

- *Supporting livelihoods-based programming:* Effective MEAL systems are essential for determining if the response is meeting the LEGS livelihoods objectives and achieving the planned impact on livestock-based livelihoods.
- *Ensuring community participation:* It is vital that MEAL systems include community participation. This will ensure the views and concerns of affected communities are heard and addressed during implementation, while promoting accountability. Participation also helps to ensure that monitoring data is reliable. It is the livestock keepers themselves who are best placed to observe the impacts of interventions over time.

LEGS framework
LEGS Principles

- *Supporting gender-sensitive programming:* Disaggregating MEAL data helps to identify the potentially different needs of men and women. It also shows whether a planned response is reaching or impacting them differently.
- *Supporting local ownership:* Involving affected communities in monitoring and evaluating the emergency response has the potential to increase their engagement in its implementation. Communities should be involved right from the identification of initial indicators to the data collection process itself.

Appendices

Appendix 2.1: LEGS alignment with the Sphere Protection Principles

In many parts of the world, livestock are valuable financial assets and a ready source of high-quality food. Livestock are also mobile. Therefore, in insecure environments, looters and armed groups may target livestock. To ensure the protection of people involved in livestock-related emergency responses and to minimise risk, proper analysis of protection issues prior to intervention is needed. For example:

- The protection or distribution of livestock may increase individual household vulnerability to theft or looting as a deliberate tactic of war. The extent to which livestock are an asset rather than a liability depends on the particular security context.
- Livestock management may require women and children to travel to remote areas to find feed or water for animals. This can place them at risk of violence, sexual abuse or abduction.
- Displaced people may be particularly at risk. Concentration of livestock may attract theft, or travelling through unfamiliar areas for water or grazing may increase vulnerability to attack.
- In times of natural resource scarcity, the movement of livestock to new areas can increase the potential for conflict between host and visiting communities.
- As a result of crisis, children may become the main owners of livestock. This may put them at risk of theft and violence. It is important to ensure that livestock interventions are safe for vulnerable children and ensure their access to age-specific, dignified support.

This type of analysis should form part of the initial assessment *(see Chapter 3: Emergency response planning)*, especially in conflict-related emergencies.

LEGS framework
LEGS Principles

Appendix 2.2: How LEGS Principles support the Core Humanitarian Standard Commitments

Core Humanitarian Standard Commitments and quality criteria	How LEGS Principles support the CHS Commitments
1. Communities and people affected by crisis receive assistance appropriate and relevant to their needs. Quality criterion: Humanitarian response is appropriate and relevant.	If participatory approaches are used properly during initial assessment and response identification, assistance is more likely to be appropriate and relevant. See LEGS *Principle 2: Ensuring community participation*; LEGS *Chapter 3: Emergency response planning*.
2. Communities and people affected by crisis have access to the humanitarian assistance they need at the right time. Quality criterion: Humanitarian response is effective and timely.	LEGS *Principle 4: Preparedness and early action*, supports early and timely response. Use of the LEGS Participatory Response Identification Matrix supports appropriate responses against different stages of an emergency. LEGS impact indicators include the timeliness of response.
3. Communities and people affected by crisis are not negatively affected and are more prepared, resilient and less at risk as a result of humanitarian action. Quality criterion: Humanitarian response strengthens local capacities and avoids negative effects.	LEGS *Principle 1: Livelihoods-based programming* and *Principle 7: Supporting local ownership* recommend working with local actors, systems and services where possible.
4. Communities and people affected by crisis know their rights and entitlements, have access to information and participate in decisions that affect them. Quality criterion: Humanitarian response is based on communication, participation and feedback.	LEGS *Principle 2: Ensuring community participation* emphasises community involvement in design, implementation and evaluation of interventions. LEGS *Chapter 3* emphasises community participation throughout the planning of the response.

continued over

LEGS framework
LEGS Principles

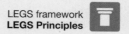

Core Humanitarian Standard Commitments and quality criteria	How LEGS Principles support the CHS Commitments
5. Communities and people affected by crisis have access to safe and responsive mechanisms to handle complaints. Quality criterion: Complaints are welcomed and addressed.	Complaint mechanisms should be generic across an organisation and are not livestock-specific. .... LEGS *Principle 8: MEAL*, emphasises participatory monitoring as well as upward accountability.
6. Communities and people affected by crisis receive coordinated, complementary assistance. Quality criterion: Humanitarian response is coordinated and complementary.	LEGS *Principle 5: Coordinated responses*, builds on a strong evidence base showing links between good coordination and impacts.
7. Communities and people affected by crisis can expect delivery of improved assistance as organisations learn from experience and reflection. Quality criterion: Humanitarian actors continuously learn and improve.	LEGS *Principle 8: MEAL*, stresses the importance of learning from monitoring and evaluation to inform future planning. .... Producing new editions of LEGS includes systematic evidence reviews, including reference to participatory impact evaluations.
8. Communities and people affected by crisis receive the assistance they require from competent and well-managed staff and volunteers. Quality criterion: Staff are supported to do their job effectively, and are treated fairly and equitably.	LEGS *Principle 4: Preparedness and early action*, highlights the importance of ensuring that staff have the necessary technical experience and skills in livelihoods-based and participatory approaches. .... LEGS *Principle 6: Gender-sensitive programming*, recognises that the gender of staff will be important in certain contexts.

continued over

LEGS framework
LEGS Principles

Core Humanitarian Standard Commitments and quality criteria	How LEGS Principles support the CHS Commitments
9. Communities and people affected by crisis can expect that the organisations assisting them are managing resources effectively, efficiently and ethically. Quality criterion: Resources are managed and used responsibly for their intended purpose.	LEGS supports appropriate resource management and channelling resources to the right type of livestock response at the right time. Coordinated responses lead to efficient use of resources; see LEGS *Principle 5: Ensuring coordinated responses*.

References and further reading

LEGS livelihood principle

Aklilu, Y., Admassu, B., Abebe, D. and Catley, A. (2006) *Pastoralist livelihood initiative: guidelines for livelihoods-based livestock relief interventions in pastoralist areas*, Feinstein International Famine Center, Tufts University, Addis Ababa, https://www.livestock-emergency.net/userfiles/file/general/Feinstein-International-Center-2006.pdf

Braam D.H., Chandio R., Jephcott F.L., Tasker A., Wood J.L.N. (2021) Disaster displacement and zoonotic disease dynamics: The impact of structural and chronic drivers in Sindh, Pakistan. *PLOS Glob Public Health* 1(12): e0000068, https://doi.org/10.1371/journal.pgph.0000068

Schreuder, B. (2015) *Afghanistan, a 25-years' Struggle for a Better Life for its People and Livestock*, DCA-VET/Erasmus Publishing, Lelystad/Rotterdam, https://dca-livestock.org/books/

Scoones, I. (1998) *Sustainable rural livelihoods: a framework for analysis*, IDS working paper 72, Institute for Development Studies, University of Sussex, Brighton, https://opendocs.ids.ac.uk/opendocs/bitstream/handle/20.500.12413/3390/Wp72.pdf?sequence=1

LEGS community participation principle

Catley, A. and Leyland, T. (2001) 'Community participation and the delivery of veterinary services in Africa', *Preventive Veterinary Medicine* 49: 95–113, https://doi.org/10.1016/S0167-5877(01)00171-4

Catley, A. (2020) Participatory epidemiology: reviewing experiences with contexts and actions, *Preventive Veterinary Medicine* 180, 105026, https://doi.org/10.1016/j.prevetmed.2020.105026

Catley, A., Leyland, T. and Bishop, S. (2008) 'Policies, Practice and Participation in Protracted Crises: The Case of Livestock Interventions in South Sudan', in L. Alinovi, G. Hemrich and L. Russo (eds), *Beyond Relief: Food Security in Protracted Crises*, Practical Action Publishing, Rugby, https://www.researchgate.net/publication/257825324_Alinovi_L_Hemrich_G_Russo_L_editors_Beyond_Relief_Food_Security_in_Protracted_Crises_Evidence_from_Sudan_Somalia_and_the_Democratic_Republic_of_Congo_Practical_Action_London_2008

CBM Global Disability Inclusion (no date) *Step-by-step practical guidance on inclusive disaster risk reduction*, Inclusive DRR Hands-on Tool, CBM Global, https://idrr.cbm.org/en/card/livestock

Cornwall, A. (1996) 'Towards participatory practice: participatory rural appraisal (PRA) and the participatory process', in K. de Koning, M. Martin (eds), *Participatory Research and Health: Issues and Experiences*, Zed Books Ltd., London, pp.94–107, https://www.eldis.org/document/A15874

Sadler, K., Mitchard, E., Abduallahi A., Shiferaw Y. and Catley, A. (2012) *Milk Matters: The Impact of Dry Season Livestock Support on Milk Supply and Child Nutrition in Somali Region, Ethiopia*, Feinstein International Center, Tufts University, Addis Ababa, https://fic.tufts.edu/wp-content/uploads/Milk-Matters-2.pdf

LEGS climate and environment principle

Houzer, E. and Scoones, I. (2021) *Are livestock always bad for the planet? Rethinking the protein transition and climate change debate*, Brighton: PASTRES, Brighton, https://opendocs.ids.ac.uk/opendocs/bitstream/handle/20.500.12413/16839/Climate-livestock_full_report_%28EN%29_web.pdf?sequence=5&isAllowed=y

VSF International (2014) *Livestock and climate change – going beyond preconceived ideas and recognizing the contribution of small-scale livestock farming facing climate change*, Position Paper n.2. Executive Summary, October 2014, http://vsf-international.org/wp-content/uploads/2015/02/2_LivestockClimChange_english-new-logo.pdf

LEGS framework
LEGS Principles

LEGS preparedness and early action principle

Benson, C. and Twigg, J. with Rossetto, T. (2007) *Tools for Mainstreaming Disaster Risk Reduction: Guidance Notes for Development Organisations*, ProVention Consortium, Geneva, https://www.preventionweb.net/files/1066_toolsformainstreamingDRR.pdf

Cabot Venton, C., Fitzgibbon, C., Shitarek, T., Coulter, L. and Dooley, O. (2012) *The economics of early response and disaster resilience: lessons from Kenya and Ethiopia*, DFID, London, https://www.gov.uk/government/uploads/system/uploads/attachment_data/file/67330/Econ-Ear-Rec-Res-Full-Report_20.pdf

Catley, A. and Charters, R. (2015) *Early response to drought in pastoralist areas: lessons from the USAID crisis modifier in East Africa*, USAID/East Resilience Learning Project, USAID/East Africa, Nairobi, https://pdf.usaid.gov/pdf_docs/PA00M1PX.pdf

Catley, A. and Cullis, A. (2012) 'Money to Burn? Comparing the costs and benefits of drought responses in pastoralist areas of Ethiopia', *Journal of Humanitarian Assistance*, 24 April, 2012, https://www.livestock-emergency.net/wp-content/uploads/2020/05/Catley-and-Cullis-2012.pdf

LEGS (2018) *Revisiting the economic impacts of early drought response: How does early response affect households in pastoralist areas?* LEGS Briefing Paper, Livestock Emergency Guidelines and Standards, https://www.livestock-emergency.net/wp-content/uploads/2017/10/LEGS-Briefing-Paper-Economic-Impacts-of-Early-Drought-Response.pdf

Rohwerder, B. (2017) *Flexibility in funding mechanisms to respond to shocks*, GSDRC Helpdesk Research Report 1412, Birmingham, UK: GSDRC, University of Birmingham, https://assets.publishing.service.gov.uk/media/59b7ec37ed915d19636fef39/1412-Flexibility-in-funding-mechanisms-to-respond-to-shocks.pdf

USAID/Ethiopia (2014) *Resilience in Africa's drylands: revisiting the drought cycle management model*, USAID/Ethiopia Agriculture Knowledge, Learning, Documentation and Policy Project, Technical Brief, May 2014, http://www.agri-learning-ethiopia.org/wp-content/uploads/2014/07/AKLDP-Technical-Brief-DCM-May-2014-HQ.pdf

LEGS coordination principle

Watson, D.J. and van Binsbergen, J. (2008) *Review of VSF-Belgium's Turkana emergency livestock off-take intervention* 2005, Research Report 4, International Livestock Research Institute (ILRI), Nairobi, https://cgspace.cgiar.org/bitstream/handle/10568/234/RR4_LivestockOfftakeIntervention.pdf

LEGS gender-sensitive programming principle

de Jonge, K. and Maarse, L. (2020) *Gender and livestock in emergencies: a discussion paper for the Livestock Emergency Guidelines and Standards (LEGS)*, LEGS, https://www.livestock-emergency.net/wp-content/uploads/2020/11/LEGS-Discussion-Paper-Gender-and-Livestock.pdf (See for example Box 1: 'Five Steps in a Rapid Gender Analysis'.)

IFAD (2020) *How to do gender and pastoralism*, International Fund for Agricultural Development, Rome, https://www.ifad.org/en/web/knowledge/-/publication/how-to-do-note-gender-and-pastoralism

LEGS (2018) *LEGS gender module*, On-line training module on LEGS and gender, https://www.livestock-emergency.net/new-line-training-module-legs-gender/

Mazurana, D. and Maxwell, D. (2016) *Sweden's feminist foreign policy: implications for humanitarian response*, FIC/Tufts, HPG, Kings College London, World Peace Foundation, https://fic.tufts.edu/assets/Feminist-Foreign-Policy-Humanitarian-Response.pdf

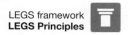

Mercy Corps (2014) *Rethinking resilience: prioritizing gender integration to enhance household and community resilience to food insecurity in the Sahel*, https://www.mercycorps.org/sites/default/files/2019-12/Rethinking_Resilience_Gender_Integration.pdf

Pasteur, K. (2002) *Gender analysis for sustainable livelihoods frameworks: tools and links to other sources*, draft, https://www.livestock-emergency.net/userfiles/file/assessment-review/Pasteur-2002.pdf

UNFPA (2015) *Minimum Standards for Prevention and Response to Gender-based Violence in Emergencies*, United Nations Population Fund (UNFPA), https://www.unfpa.org/publications/minimum-standards-prevention-and-response-gender-based-violence-emergencies-0

LEGS supporting local ownership principle

Featherstone A. (2020) *Putting the best foot forward: localisation, contextualisation and institutionalisation, a discussion paper for the Livestock Emergency Guidelines and Standards (LEGS)*, LEGS, https://www.livestock-emergency.net/wp-content/uploads/2020/11/LEGS-Institutionalisation-and-Localisation-Summary-Discussion-Paper.pdf

Humanitarian Leadership Academy (2019) *Unpacking localization*, https://www.humanitarianlibrary.org/resource/unpacking-localisation

Humanitarian Practice Network (2021) *Survivor- and community-led crisis response: practical experience and learning*, Briefing Paper no. 84, https://odihpn.org/wp-content/uploads/2021/05/HPN_SCLR-Network-Paper_WEB.pdf

OECD (2017) *Localising the response: World Humanitarian Summit – putting policy into practice*, The Commitments into Action Series, OECD, Paris, https://www.oecd.org/development/humanitarian-donors/docs/Localisingtheresponse.pdf

See also: Local to Global Protection, https://www.local2global.info/

LEGS MEAL principle

FAO (2016) *Livestock-related interventions during emergencies – The how-to-do-it manual.* Edited by Philippe Ankers, Suzan Bishop, Simon Mack and Klaas Dietze. FAO Animal Production and Health Manual No. 18. Rome, https://www.fao.org/3/i5904e/i5904e.pdf (See Chapter 10, Monitoring, Evaluation and Impact Assessment)

Sphere Association (2018) *The Sphere Handbook: Humanitarian Charter and Minimum Standards in Humanitarian Response*, The Sphere Project, Geneva, https://spherestandards.org/handbook-2018/

See also case studies on LEGS Principles at: https://www.livestock-emergency.net/resources/case-studies/

LEGS framework
Emergency response planning

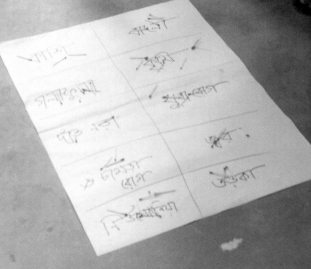

Chapter 3: Emergency response planning

78	**Introduction**
79	**Stage 1: Initial assessment**
87	**Stage 2: Response identification**
90	**Stage 3: Analysis of technical interventions and options**
96	**Stage 4: Response plan**
98	**MEAL guidance for LEGS interventions**
105	**Appendix 3.1: Initial assessment – suggested participatory methods**
109	**Appendix 3.2: Initial assessment – example checklist**
111	**Appendix 3.3: Initial assessment – data collection template**
112	**Appendix 3.4: PRIM – templates**
114	**Appendix 3.5: PRIM – samples**
119	**Appendix 3.6: Cash and voucher assistance – response modalities, delivery mechanisms and decision tree**
124	**Appendix 3.7: Response plan – template**
125	**Appendix 3.8: Response plan – example of livestock offtake intervention**
126	**References and further reading**

Photo © David Hadrill

LEGS framework
Emergency response planning

Chapter 3: Emergency response planning

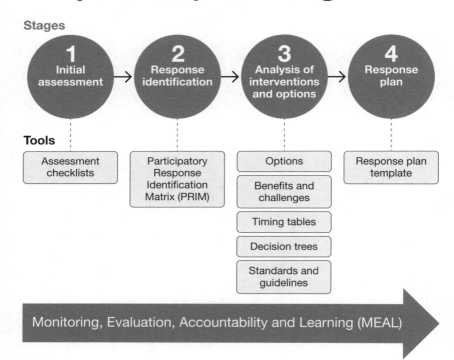

Introduction

This chapter provides LEGS guidance on emergency response planning. There are four stages in this process, each with tools to facilitate the planning process and guide the development of a response plan:

Stage 1: Initial assessment – to decide if livestock support is appropriate for a given emergency; the initial assessment generates information on livestock roles, the impact of the emergency, and an analysis of the current situation. This information is used to inform Stage 2 and Stage 3.

- Tools for Stage 1: Assessment checklists *(see also Appendix 3.2 and Appendix 3.3)*.

Stage 2: Response identification – to identify and prioritise technical interventions for livestock-based responses that are relevant and timely:

- Tools for Stage 2: Participatory Response Identification Matrix *(see also Appendix 3.4* and *Appendix 3.5)*.

Stage 3: Analysis of technical interventions and options – to select the appropriate, feasible and timely options within the prioritised intervention areas:

- Tools for Stage 3: Options, benefits and challenges tables, timing tables, decision trees, standards and guidelines *(see technical standards chapters 4, 5, 6, 7, 8, 9,* and *Appendix 3.6)*.

Stage 4: Response plan – to use the information and decisions from Stages 1 to 3 to design a response plan:

- Tools for Stage 4: Response plan template *(see Appendix 3.7* and *Appendix 3.8)*.

The four stages are all supported by **Monitoring, Evaluation, Accountability and Learning (MEAL)**, as defined in *Principle 8 (*see Chapter *2)*. MEAL systems should be established as soon as possible during planning to ensure that effective MEAL is carried out throughout the implementation of the response.

Stage 1: Initial assessment

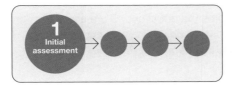

Overview

An initial assessment is needed before any emergency response, to determine **whether livestock-based support is appropriate** in the specific context. The type, phase and severity of the emergency will determine the context. This initial assessment is not an end in itself, but the first step to enable decisions to be made about which technical interventions to explore, or whether a livestock response is necessary at all.

LEGS framework
Emergency response planning

Preparing for the assessment
Review of existing information

Some of the assessment information may already have been collected before the onset of the emergency as part of preparedness planning (see *Principle 4* in *Chapter 2)*. Chapters 4–8 of the handbook each include a standard on preparedness to support this process. Agencies already working in an area are often well placed to develop preparedness capacity together with local communities. This preparedness information can provide an understanding of livelihood strategies; production systems; social, cultural and gender norms; and key stakeholders and institutions. This will significantly increase the accuracy of the initial assessment.

Secondary data can be compiled from government reports, health and veterinary statistics, NGO reports, and other sources. Spatial data from satellite photographs and geographic information systems (GIS) may also be useful. For example, it can show the extent of flooding or locations of water points and other natural resources essential for livestock.

The review of existing information is important both to avoid unnecessary demands being made on crisis-affected communities and in settings where access to conduct initial assessments is difficult.

Assessment team and methodologies

The LEGS initial assessment should be undertaken as part of a participatory planning process involving key stakeholders, including representatives of key groups within affected communities and local government. This allows an assessment of the capacities and knowledge of affected communities to respond to the emergency. It also ensures that local actors are equal partners and that they are given autonomy in designing and implementing the response plan. (See *Principle 2: Participation* and *Principle 7: Local ownership* in *Chapter 2*.)

The assessment team should therefore include community representatives and involve local government and non-governmental institutions as partners. It is important that the team is gender-balanced, that marginalised groups are represented and that it includes both generalists and livestock specialists with local knowledge.

The quality of the assessment information depends on the skills of the assessment team in participatory methods and approaches, as well as on their technical knowledge. LEGS recommends qualitative, participatory information collection for the initial assessment wherever possible. The section on MEAL at the end of this chapter provides more information on participatory approaches. *Appendix 3.1* then suggests which methods may be most appropriate for the initial assessment. Assessment teams may therefore need to be trained in appropriate participatory methodologies.

The assessment should ensure proper coverage of different groups, in particular those at risk. It should also disaggregate the findings accordingly (for example, according to age, wealth and gender) so that the needs and priorities of the different groups are addressed in the response plan.

Tools for Stage 1

The LEGS initial assessment comprises **three checklists of questions** *(see Figure 3.1 and Boxes 3.1–3.3)*. These can be answered quite rapidly using participatory methods supplemented by any existing information. The three checklists can be applied at the same time, and the questions are not fixed but can be adapted to suit the context.

Each checklist ends with a 'decision point', which helps to determine if a livestock-based response is appropriate in this context. *Appendix 3.2* presents an example of a partially completed checklist.

Figure 3.1: Summary of LEGS initial assessment checklists

1. The role of livestock in livelihoods

How do livestock contribute to livelihoods in 'normal' times?

2. Nature and impact of the emergency

How has the emergency affected communities, their livestock, and livestock management?

3. Situation analysis

What is the context (communications, infrastructure, security, other stakeholders)?

LEGS framework
Emergency response planning

> **Box 3.1**
> # Checklist 1: The role of livestock in livelihoods
>
> Livestock support is most likely to be needed if livestock are important in the livelihoods of the people affected by the emergency. The following questions help the assessment team to determine the significance of livestock in livelihoods and the role they play. They therefore help decide whether a livestock-related response is appropriate. Much of this information may already be available from existing sources. For mobile or less accessible communities in emergency contexts, responses to these questions are likely to depend largely on such pre-existing information. Whatever the source of the information, it is important to understand how livestock are managed. It is also important for the team to know how the benefits, ownership and care of livestock are affected by factors such as gender, wealth or at-risk group.
>
> 1.1 What are the main livelihood strategies in the affected area in 'normal', non-emergency, times?
>
> 1.2 What are the key uses of livestock (for example, food, income, social, draught power, transport)?
>
> 1.3 What percentage of food is derived from livestock in 'normal' times and by season?
>
> 1.4 What percentage of income is derived from livestock in 'normal' times (including livestock products, transport and draught power), and how is it managed?
>
> 1.5 What roles do different household members play in livestock care and management (including use and disposal rights)? Pay particular attention to their gender and age. Take note too of different livestock species and ages as well as seasonal variations.
>
> 1.6 What customary or other institutions and leaders are involved in livestock production and natural resource management, and what are their roles?
>
> 1.7 What are the main coping strategies and indicators for difficult times? For example, use of famine foods; high livestock slaughter or sales; abnormal migration; dispersal of household members; sale of other assets. Do these strategies have negative implications for future livelihood security?

**LEGS framework
Emergency response planning**

▶ *Decision point: Do livestock play a significant role in the livelihoods of the affected people, and is a livestock-related response therefore appropriate?*

Box 3.2
Checklist 2: The nature and impact of the emergency

This checklist focuses on understanding the impact of the emergency on the affected populations. It determines whether an emergency response is necessary. Some of this information may already be available from other agencies' assessments or, in the case of protracted emergencies, from earlier assessments.

2.1 What type of emergency is it: rapid-onset, slow-onset or complex?

2.2 What is the cause of the emergency (drought, flood, earthquake, conflict, etc.), and what area does it cover?

2.3 What is the history of this type of emergency in this area?

2.4 Which stage has the emergency reached (**alert/alarm/emergency/recovery** for slow-onset emergencies; or **immediate aftermath/early recovery/recovery** for rapid-onset emergencies)?

2.5 What human and livestock populations are affected?

2.6 What has been the impact of the emergency on the affected population? Specifically:

   2.6.1 What is the nutritional status of the affected human population?

   2.6.2 What is the prevalence of disease?

   2.6.3 What is the mortality rate?

   2.6.4 What has been the impact on different groups – men, women, children, older people, people with disabilities, particular ethnic or other social groups?

   2.6.5 What is the capacity of the affected community to respond to the emergency?

   2.6.6 Are there signs that the coping strategies and indicators for difficult times from question 1.7 above are being used?

LEGS framework
Emergency response planning

- **2.6.7** Has there been significant migration or displacement of parts of the affected population? If so, who is affected, and have they taken their livestock with them?
- **2.6.8** If the affected community has been displaced, what is the impact of this on the host community?

2.7 What has been the impact of the emergency on livestock? Differentiate by species if appropriate. Specifically:

- **2.7.1** How has livestock condition deteriorated?
- **2.7.2** What is the impact on livestock welfare (for example, lack of feed and water; injuries; disease; extreme cold; or heat stress)?
- **2.7.3** Has livestock productivity been affected (for example, offtake of milk, blood, eggs, draught power, etc.)?
- **2.7.4** Has livestock morbidity increased?
- **2.7.5** Has livestock slaughter for home consumption increased?
- **2.7.6** What is the scale of livestock losses relative to 'normal' times?
- **2.7.7** Has there been any impact on livestock shelter/enclosures?

2.8 What has been the impact of the emergency on livestock management strategies? Specifically:

- **2.8.1** What is the impact on access to grazing and/or feed?
- **2.8.2** What is the impact on access to water for livestock?
- **2.8.3** What is the impact on daily and seasonal movements?
- **2.8.4** What is the impact on livestock traders and key livestock input and output markets (sales; prices; terms of trade between livestock and cereals; feed and drug suppliers)?
- **2.8.5** What is the impact on livestock services such as veterinary services, extension services and veterinary pharmacies?
- **2.8.6** What has been the impact on the gender division of labour?

2.9 How has the environment been impacted by the emergency, and what are the implications for livestock management?

LEGS framework
Emergency response planning

2.10 What weather is forecast and what are the weather trends for the forthcoming season (anticipated snow, rains, heat, dry season, increasing insecurity, access to food, etc.)?

▶ *Decision point: Is an emergency intervention necessary?*

Box 3.3
Checklist 3: Situation analysis

This checklist helps to ensure an understanding of the operating environment, potential logistical constraints, and overlap or potential complementarity with other stakeholders.

3.1 What is the history of emergency response in the affected area (both positive and negative), and what are the lessons learned from it?

3.2 Who are the key stakeholders in the affected area, including local responders, and what are they doing?

3.3 Are any stakeholders playing a coordination role, and how effective is the coordination?

3.4 What resources are available, especially customary coping strategies and locally led response?

3.5 What is the current context?

   3.5.1 How are communications functioning?

   3.5.2 What is the security situation, and what are the implications for programming and staff safety?

   3.5.3 What are the implications for livestock movement and migration (rights of access, potential conflict)?

   3.5.4 What are the key protection issues facing livestock keepers?

   3.5.5 What is the current infrastructure, such as roads and transport?

3.5.6 What is the context for potential cash and voucher assistance (CVA)? For example, are markets accessible and functioning? Is CVA accepted by recipients and the government? Are there secure and regulated delivery options such as mobile phones, remittance companies, etc.?

3.5.7 Are there any cross-border issues?

3.5.8 What are the policy and/or legal challenges affecting livestock-related interventions? For example, livestock movement or export bans; slaughter laws; licensing regulations (relating to community-based animal health workers, for instance); coordination between aid organisations; and organisational policies of key stakeholders.

3.5.9 Are particular interests (cultural, political, etc.) likely to have an impact on potential interventions?

These questions become especially significant in conflict situations.

▶ *Decision point: Do answers to any of the above constitute critical issues that prevent any form of intervention in the area? For example, does the security situation hinder any kind of movement at present? Are other actors already providing sufficient support to affected populations?*

Recording and analysing the information

The assessment team should record the results of the initial assessment using simple templates, showing the findings, methods used, sources and dates (see *Appendix 3.3* for an example recording template). These can then be compiled into a report, structured around the questions in the checklists. The team should then meet to review and analyse the results, reflecting on each checklist's key findings and decision points in order to decide if a livestock-based response is appropriate.

Stage 2: Response identification

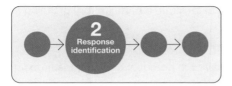

Overview

If the initial assessment concludes that livestock support may be appropriate, the next stage is to identify **which livestock-based emergency responses would be *relevant* and *timely*.** The assessment team should also design livestock responses in emergencies to meet at least one of the LEGS livelihoods objectives. That is, **support crisis-affected communities to:**

1. **obtain immediate benefits** using existing livestock assets; and/or
2. **protect key livestock assets**; and/or
3. **rebuild key livestock assets**.

LEGS presents six technical interventions that can provide livestock-based livelihood support to affected communities in an emergency. They may be used at the same time and may complement each other. Examples include providing both feed and water to drought-affected livestock; or veterinary support alongside provision of livestock to replace animals lost in an earthquake. The six interventions are:

- **feed:** animal feed or fodder to crisis-affected animals;
- **water:** water for livestock through support to existing supplies or creation of new sources;
- **veterinary support:** animal health services and support;
- **shelter:** protection of livestock affected by extreme weather conditions, insecurity or displacement;
- **livestock offtake:** removal (through sale or slaughter programmes) of crisis-affected animals before they lose their market or food value – livestock offtake is generally only applicable in drought emergencies; this differs from disease outbreaks, when carcass disposal is necessary;
- **provision of livestock:** replacement of livestock lost in the emergency, or provision of animals to support new livelihood activities after an emergency.

LEGS framework
Emergency response planning

Tool for Stage 2

The LEGS tool for Stage 2 is the **Participatory Response Identification Matrix (PRIM)**. The aim of the PRIM is to facilitate a participatory planning process that:

- provides rapid and highly visual results;
- demonstrates which interventions have the potential to have most impact;
- focuses on the purpose (livelihoods objective) of any response; and
- confirms the appropriate timing.

The PRIM is best implemented in a workshop setting. Where a face-to-face meeting is not possible or appropriate, the team may use alternative methods such as online meetings. The participants for the PRIM process should involve all key stakeholders. These include local leadership, representatives of the affected populations (both women and men) and representatives of key at-risk groups. This will facilitate broad participation in (and local ownership of) the response planning process. In turn, this will help to identify local knowledge and skills, and recognise local capacities to address and respond to the emergency.

In the left side of the PRIM, the six technical intervention areas (feed; water; veterinary support; shelter; livestock offtake; and provision of livestock) should be considered against the three LEGS livelihoods objectives. This allows workshop participants to review how much each intervention could impact on each objective *(see Figure 3.2)*.

The right side of the matrix should show the phases of the current emergency and note the best timing of each intervention.

LEGS provides two PRIM templates, one for rapid-onset emergencies and the other for slow-onset emergencies. These reflect the two sets of definitions of the emergency phases outlined in *Chapter 1 (*see the blank PRIM templates in *Appendix 3.4)*. For complex emergencies, workshop participants can use the PRIM that relates most closely to the current context. Otherwise, they can agree on their own definitions and insert them into the right-hand side of the matrix.

LEGS framework
Emergency response planning

Figure 3.2: How to complete the Participatory Response Identification Matrix

1. Select the appropriate PRIM format according to the type of emergency (rapid-onset or slow-onset) and confirm that the emergency phases are appropriate to the current context (see *Chapter 1: Introduction to LEGS* on *How does LEGS define emergencies and what are their impacts?*). Note that the headings on the right side of the PRIM vary according to the type of emergency (rapid-onset or slow-onset). The example in Figure 3.2 is for a rapid-onset emergency.

2. Consider each of the potential technical interventions against the three LEGS livelihoods objectives. See *Table 1.1* in Chapter 1 and the introduction to each of the technical chapters (4–9). Here there is information on how each technical intervention may contribute to the LEGS livelihoods objectives.

LEGS framework
Emergency response planning

3. Provide scores against each objective to show how much the technical intervention has the potential to impact on that objective. Score each from 0 to 3, where 3 = very positive impact on the objective; 2 = some impact; 1 = very little impact; 0 = not applicable.
4. Add ticks (check marks) to show the optimum timing of each intervention.
5. Review the matrix, and note which interventions have the most potential for positive impact given the timing of the emergency. Note that the LEGS technical interventions are not exclusive and that more than one may be prioritised. Different interventions may also be implemented together or one after the other over the course of the emergency.

There is no universally 'correct' PRIM; each PRIM is developed by participants based on their specific location and needs. Participants should also be aware of potential biases based on individuals' personal interests or expertise when completing the matrix.

Blank PRIM templates for both rapid-onset and slow-onset emergencies are presented in *Appendix 3.4*, while examples of completed PRIMs are presented in *Appendix 3.5*.

The output of a PRIM is agreed identification of the most **relevant** and **timely** interventions to support and protect livestock-based livelihoods in the specific context and for the current phase of the emergency. This provides the basis for moving to Stage 3.

Stage 3: Analysis of technical interventions and options

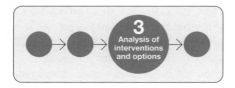

Overview

In Stage 2, workshop participants prioritise one or more technical interventions using the PRIM. In Stage 3, these selected interventions are analysed in more detail to determine which of their specific options will be **appropriate, feasible and timely.** To do this, a range of tools are provided in each of the technical chapters (4–9).

For each of the six LEGS technical interventions that were introduced in Stage 2 (for example water or livestock offtake), there is a choice of options for delivering the intervention. Examples include water trucking versus rehabilitation of water pumps; or commercial offtake versus slaughter offtake, as shown in *Table 3.1*.

Table 3.1: Summary of technical interventions and options

Technical intervention	Technical options
Feed	Home-based emergency feeding
	Feed camp emergency feeding
Water	Water points
	Water trucking
Veterinary support	Clinical veterinary services
	Public sector veterinary functions
Shelter	Livestock shelter
	Livestock and settlement
Livestock offtake	Commercial offtake
	Slaughter offtake for consumption
Provision of livestock	Replacing lost livestock assets
	Building livelihood assets

LEGS framework
Emergency response planning

Tools for Stage 3

Each technical chapter of LEGS contains **five key tools** for analysing the suitability and feasibility of the selected intervention(s) and option(s). These tools are designed to be used as needed. They do not necessarily have to be followed in order, and not all have to be used. A suggested process to follow is:

1. Review the section on **options** at the start of the relevant technical chapter to prioritise the most appropriate option(s).
2. Review the **benefits and challenges table** to confirm the choice of option(s).
3. Review the **timing of interventions table** to confirm that the intervention is appropriate for the emergency phase for which it is planned.
4. Work through the **decision tree** to check the context and requirements for the selected intervention/option.
5. Review the **standards, key actions and guidance notes** to plan in more detail.

Tool 1: Outline of options

Each of the six technical chapters provides an explanation of the different options possible within that technical intervention. For example, the feed chapter presents two options for emergency feeding: (a) home-based; and (b) feed camp. Feed option (b) is then divided into two sub-options: in-out (daily) feed camp, and residential feed camp.

Begin by reviewing this explanation to identify which option(s) are most appropriate for the context.

Tool 2: Benefits and challenges table for the different options

Each technical chapter has a table that summarises the benefits and challenges of the options (see, for example, Table 4.3 in Chapter 4: Livestock feed). Review this table to confirm the appropriate selection of the option(s).

Tool 3: Timing table

Each technical chapter has a table showing the most appropriate timing for the various options according to the phases of the emergency for both rapid-onset and slow-onset emergencies (see, for example, Table 4.4 in Chapter 4: Livestock feed). These timings are suggestions only, as the situation may vary according to local conditions.

Review the table for guidance on the optimum timing for the selected option(s) according to the context. *Annex C* (at the end of the handbook) presents a combined table for all the timing tables in the technical chapters.

Tool 4: Decision tree

The decision tree in each technical chapter brings together some of the key questions and issues to consider before a particular technical option can be decided upon *(see, for example, Figure 4.1* in *Chapter 4: Livestock feed).*

Work through the decision tree questions following the 'yes' or 'no' arrows to arrive at one or more of the technical options – or at a 'no action' box. The information from the initial assessment and PRIM process will help to provide answers to these questions. The result 'no action' may mean additional preparation is necessary to be able to answer 'yes' to the key questions, rather than that no intervention should take place.

Tool 5: Standards and guidelines

Each technical chapter contains standards, key actions and guidance notes as the main content of the chapter. They provide more information and issues to consider in designing a response, and are defined as follows:

- **Standards are statements describing the minimum to be achieved in any emergency in any context and are generally qualitative.**
- **The key actions attached to each standard are key steps or actions that contribute to achieving the standard.**
- **Guidance notes, which should be read in conjunction with the key actions, outline particular issues to consider when applying the standards.**

The technical chapters also provide a summary of how the eight LEGS Principles relate to each intervention, as well as specific issues to consider when undertaking the intervention.

All the technical chapters include a standard on **assessment and planning** (as well as an assessment checklist in the chapter appendices). This standard provides key information and questions that implementing agencies need to address in order to develop the response plan. Most of the technical chapters also include a standard on **preparedness,** which outlines key actions and preparations implementing agencies can undertake prior to the onset of an emergency. These activities should inform the planning process and, if appropriate, be incorporated into the response plan.

The other standards, key actions and guidance notes in the technical chapters present information on particular aspects of the intervention or option. They also provide detailed technical guidance to support the design and implementation of the response plan.

Delivery modalities and mechanisms

As part of the preparation for the response plan, the implementing agency should review the appropriate modality and delivery mechanism for the response.

Modalities refers to the way assistance is delivered. This can be either in-kind assistance, where goods or services are provided directly to recipients; or cash and voucher assistance (CVA), where recipients are provided with either cash or vouchers to purchase goods or services of their choice. CVA is increasingly used in humanitarian response as it allows recipients of assistance to set their own priorities and make choices. There is growing evidence showing CVA's positive impact in protecting or restarting livelihoods.

The implementing agency's choice of modality depends on several factors. These include the market availability (or potential market availability) of the goods and services required, recipients' preferences for cash, vouchers and/or in kind, and their ability to access the assistance. The agency also needs to consider the security and protection situation, and donor and agency requirements and capacities. The initial assessment can help determine whether CVA is feasible *(see Assessment checklist 3, question 3.5.6)*.

A **delivery mechanism** in humanitarian CVA is a means of delivering or transferring cash or vouchers to recipients (for example, smart card, mobile money transfer, over the counter, a cheque, automated teller machine/ATM card, etc.). Some delivery mechanisms may also facilitate receipt, storage and payments (for example, mobile wallet, bank account, smart card, etc.).

Appendix 3.6 presents some of the potential CVA response modalities and delivery mechanisms that may be used in LEGS technical interventions *(see Figure 3.8 and Table 3.5)*. These are followed by a decision tree that can help agencies identify if CVA is feasible and appropriate *(see Figure 3.9)*.

The choice of using cash (physical or electronic) or vouchers (paper or electronic) will also depend on the objectives of the intervention:

- Cash may be suitable where there is an immediate need (for example, to purchase feed or veterinary services). It may also be applicable when key goods and services are available in the local market.
- Vouchers may be more appropriate if there is a need to ensure quality or provision of specific goods and services (for example, if certain types of feed or veterinary services are needed, or to obtain and construct specific types of shelter).

There are also forms of indirect cash assistance, where agencies give subsidies to service providers or governments to facilitate the intervention. Examples include a waiver of slaughterhouse fees, movement permit fees, market fees, or veterinary fees; subsidised trucking costs; provision of fuel to water users' associations; or government subsidies or price caps on feed supplements.

A key consideration for CVA is the accessibility of the delivery mechanism for livestock keepers (that is, the way the assistance is conveyed to recipients). For example, delivering cash in hand or cash over the counter may not be suitable for nomadic/pastoralist livestock keepers. Here, mobile money, if available, may be more appropriate and accessible. Similarly, where an identity document (ID) is required to access CVA, there may be difficulties for livestock keepers who do not have an ID, or for displaced people whose ID is lost or not recognised.

More information on CVA, including detailed guides on market assessment and cash response mechanisms, can be found in the *References and further reading* section.

LEGS framework
Emergency response planning

Stage 4: Response plan

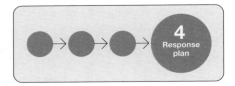

Overview

In Stage 4 of the LEGS approach to emergency response planning, the implementing agency develops a response plan. This is based on the information and decisions made in the previous three stages.

LEGS provides a template for a response plan *(see Appendix 3.7)*. However, some organisations will use their own formats and may have slightly different terminology. Whichever format (or term) is used, the response plan should be based on a well-reasoned theory of change from Activity → Output → Outcome → Objective *(see Table 3.2)*.

Table 3.2: Theory of change for a livestock response plan

Level	Description	Example – Livestock offtake
Impact	Livelihood impacts are clearly specified	At least 20% of income from livestock offtake activities is used to protect remaining livestock before the end of the response
Outcome	Short-term and medium-term positive effects of response – it has causal links with **output**	Average income derived from livestock offtake activities is US$100 per household
Output	What the response delivers to achieve the **outcome**	Up to 3 cattle per household are purchased by implementing agency
Activities	The actions to achieve each **output**	Agree maximum number of cattle to be purchased per household
		Agree target households
		Agree livestock prices
		Identify transportation, and so on…

The causal links from bottom to top of the theory of change (ToC) should be reviewed to ensure that there is a logical flow:
- **If** we do these **activities, then** this **output** will be delivered.
- **If** we deliver this **output, then** this **outcome** will be achieved.
- **If** the **outcome** is achieved, **then** this will contribute to the **objective** and **impacts.**

To ensure the technical plausibility of the flow of **activities** to **impact**, each stage of the ToC should be quantified. Along with identifying the causal links, the assumptions made in defining the theory of change and the potential risks need to be defined. These should also be monitored through the MEAL system (see section on MEAL below).

Whereas the LEGS livelihoods objectives are general, the response plan objectives should be specific, measurable, achievable, relevant and time-bound (SMART objectives). They can then quantify the intended livelihood impacts within a specific time frame. Without SMART objectives, it is difficult to determine whether an objective (and impact) has been achieved. Good-quality initial assessment and response identification in Stages 1 to 3 should facilitate this process of determining appropriate objectives.

Table 3.3: Linking LEGS livelihoods objectives to SMART objectives

LEGS livelihoods objective	Example of SMART objective in response plan
To support crisis-affected communities to obtain immediate benefits using existing livestock assets	**Slaughter offtake:** Provision of 2.2kg of dried meat covers total protein requirements of an average household of four people for at least seven days
To support crisis-affected communities to protect key livestock assets	**Supplementary feed:** Cattle mortality is reduced to less than 15% during the response in target households during project period
To support crisis-affected communities to rebuild key livestock assets	**Provision of livestock plus veterinary care:** At least 75% of restocked households build a herd of at least 40 sheep and goats by month 20 of the project

**LEGS framework
Emergency response planning**

Tool for Stage 4

The tool for Stage 4 is the response plan (see *Appendix 3.7* for a response plan template). The plan brings together the information and analysis from the previous three stages.

The response plan should show the selected technical intervention(s) and option(s), the related LEGS livelihoods objective(s), the response's SMART objectives, and the theory of change for the response. An example of a partially completed response plan for a livestock offtake intervention is presented in *Appendix 3.8*.

It is important for the implementing agency to identify process and impact indicators (see next section for definitions) as part of the response plan. Then monitoring can start as soon as implementation begins. Monitoring data is vital to ensure that activities can be adjusted during implementation. It also provides key information for later evaluation and impact assessment as part of a wider MEAL system (see below).

The response plan should ensure that interventions are implemented fairly, based on transparent and participatory targeting, and by drawing on information from the initial assessment (Stage 1) and community consultations. The implementing agency should agree targeting methods and selection of recipients with the affected communities, including representatives of at-risk groups.

The agency also needs to clearly define and disseminate selected targeting criteria. Where appropriate, public meetings can be held to increase transparency and accountability. As far as possible, the targeting process should remain in the control of recipient communities (with some oversight to ensure at-risk groups are not overlooked). This will avoid concerns about inequitable distribution of benefits and will help ensure accountability and transparency.

The completion of all four stages should result in a response plan that identifies appropriate, timely and feasible interventions and activities, based on identified need and local context, and linked to livelihoods objectives. Chapters 4–9 of the LEGS Handbook provide the technical details to support this process.

Monitoring, evaluation, accountability and learning (MEAL) guidance for LEGS interventions

Monitoring, evaluation, accountability and learning (MEAL) should support all four stages of the LEGS approach to emergency response planning. *Principle 8 (see Chapter 2)* outlines the importance of MEAL in livestock emergency response. This section provides further guidance on key aspects of MEAL that should be considered alongside the Chapter 10 'Monitoring, evaluation and impact assessment of livestock-related interventions during emergencies' in *FAO* (2016). There are also many general sources of information on MEAL *(see References and further reading)*.

Most organisations have their own MEAL systems, often led by MEAL specialists. Where these exist, the process and monitoring indicators identified in the response plan can be fed into the system so that monitoring can begin as soon as possible.

If not already set up, MEAL systems should be established as soon as possible during planning. This will ensure that monitoring can begin at the start of the initiative and that effective MEAL can be carried out throughout the implementation of the response. The MEAL system should aim to support shared learning about livelihood impacts. At a minimum, MEAL systems should:

1. use theories of change and SMART objectives;
2. define process and impact indicators;
3. combine participatory methods with other data;
4. use monitoring to revise implementation as needed;
5. include impact evaluation;
6. be accountable;
7. share and apply learning.

1. Use SMART objectives

As discussed under Stage 4, the objectives of the response should be specific, measurable, achievable, relevant, and time-bound (SMART). This helps to ensure that the response is well planned, and makes it possible to measure the impact on livelihoods.

An example of a SMART objective for a livestock feed intervention is: 'In two districts, reduce mortality by at least 25 per cent in the core small ruminant herds owned by 50 per cent of the poorest households by the end of the project'; see *FAO* (2016) Box 11, p. 164.

2. Define process and impact indicators

Indicators are items that are measured by monitoring and evaluation. LEGS uses two types of monitoring indicators: process indicators and impact indicators.

Process indicators measure the progress of the activity: for example, the number of livestock vaccinated; the number of training courses held. They are usually quantitative. For example, in an intervention to provide veterinary services, a process indicator could be the number of drugs dispensed or of veterinary visits made. Process indicators often relate to project expenditure. They can therefore be used for financial accountability, as well as to flag up the need for adjustments in implementation. Process monitoring is typically done on a regular basis (for example, monthly) throughout an intervention.

Impact indicators focus on the impact of the activity on the recipients' lives and livelihoods. Examples include improved nutrition through milk production, or increased income through livestock sales. In the veterinary services example above, an impact indicator connected to the number of treatments of animals could be livestock mortality by species and disease. Measurement of impact indicators is generally carried out as part of impact evaluation (see below).

Impact indicators usually require either an understanding of a pre-existing situation or a reference point against which impacts can be understood. For example:

- If a livestock feed or veterinary response aims to reduce mortality by 50 per cent, the mortality at the start of the intervention should be known. This baseline mortality can be discussed and agreed with communities during Stage 3.
- If a slaughter offtake response aims to provide cash to target households, information on household income or expenditure immediately before the intervention is useful. This will allow understanding of the relative importance of the cash derived from the intervention.
- If a combined livestock feed–veterinary response aims to increase children's consumption of livestock milk during a drought, information on milk consumption immediately before the intervention is useful.

Where baseline information is not already available, it may be gathered during the initial assessment or later as a 'retrospective' baseline; see pp. 165–166 (including Table 21) of *FAO* (2016).

Each of the LEGS technical chapters includes an appendix with suggested process and impact indicators. See also pp. 161 (Table 19) and 165 of *FAO* (2016) for guidance on when to measure process and when to measure impact.

3. Combine participatory methods with other data

In line with LEGS Principles, LEGS recommends participatory approaches to monitoring and evaluation. A range of well-tested participatory methods are suitable for the LEGS initial assessment and for monitoring and evaluation. They are drawn from Participatory Learning and Action (PLA), Participatory Rural Appraisal (PRA) and participatory epidemiology techniques.

Participatory methods can produce different types of information depending on the method used and other factors. For example:

- Participatory maps produce visual information that can be annotated with notes.
- Informal interviews can produce qualitative information, such as people's preferences for different livestock species or their views on the implementation of a response. These interviews can also produce quantities such as averages for daily milk production or livestock mortality, or livestock prices.
- Ranking and scoring methods show the relative importance or value of selected items, expressed as ranks or scores. See *Appendix 3.1* for more information on participatory methods.

Information from participatory methods can be cross-checked against other information. In emergency livestock responses, two important types of cross-checking are:

- During Stages 1 to 3, information such as government statistics on livestock or market activity and prices, and research reports on livestock production or other issues (if available) can support participatory assessment. Previous evaluation reports of emergency livestock responses can also provide useful information.
- During impact evaluation, process-monitoring data on response implementation should be compiled and compared with an assessment of livelihood impacts. The theory of change enables analysis of the

**LEGS framework
Emergency response planning**

plausibility of inputs leading to impacts, and this analysis is supported by the compiled monitoring data.

Other issues to consider:

- **Sampling:** Given the time constraints when planning emergency response, 'purposive samples' of representative informants/stakeholders considered critical should be identified. This can ensure, for example, that all at-risk groups are involved (for example, poor livestock keepers affected by drought, women livestock keepers, inhabitants of a flood-affected village, or working-age children – see *CPMS Pillar 4, Standards 21 and 22)*.

- **Disaggregating:** Information gathered should always be disaggregated according to gender, age and relevant at-risk groups *(see Principle 2: Community participation, Principle 6: Gender-sensitive programming* and *Chapter 1, Box 1.3: LEGS commitment to the HSP)*.

- **Numerical data:** Scoring and ranking methods produce numerical data. Repetition of these methods can produce quantitative data sets that can be summarised using conventional statistical tests.

- **Access:** In some situations, standard participatory methodologies may not be suitable or feasible. This may happen, for example, in pandemic or conflict contexts where face-to-face gatherings are inappropriate or not allowed. In such cases, alternative approaches can be used, drawing on digital tools such as mobile phones (for interviews) and, where appropriate, online meetings (for focus group discussions). When consulting children or young people, appropriate methodologies should be used with due attention to child protection standards. (See *CPMS Principle 3, Children's Participation*, and *Standard 22*, Livelihoods and child protection.)

- **Quantitative surveys** typically use questionnaire surveys and may use statistical sampling methods. In humanitarian contexts, quantitative approaches tend to be inflexible and non-participatory, and also require substantial levels of funding and technical expertise. While tools such as mobile apps can speed up data collection in quantitative surveys, the non-participatory aspect of these surveys contrasts with LEGS *Principle 2: Community participation*. Some quantitative data may be important to support monitoring against SMART objectives: for example, livestock marketing data and the cost of goods and services. However, this data can be collected using key informant interviews. (See LEGS Participatory Techniques Toolkit; and *FAO* (2016) pp. 173, 175 and 176.)

4. Use monitoring to revise implementation as needed

Monitoring should include regular dialogue with affected communities to assess the progress and performance of response implementation. This should be supported by tracking of process indicators. The theory of change developed during response plan preparation in Stage 4 above includes an analysis of risks and assumptions that may affect the flow of inputs to impacts. Community dialogue can help to track these assumptions and risks. Overall, monitoring should be used to identify issues promptly and support real-time adjustment to implementation.

5. Include impact evaluation

The LEGS Principle on MEAL emphasises the importance of evaluating emergency response to learn lessons and to prepare for future interventions. A typical evaluation assesses the achievement of project objectives, whereas impact evaluation focuses more on the livelihood, food security or nutritional changes that a response produces. LEGS recommends the use of SMART objectives that specify impacts, so 'evaluation' and 'impact evaluation' become very similar activities.

In addition to focusing on whether SMART objectives have been met, an evaluation may also include a benefit-cost analysis. This will assess the financial costs of the response against the benefits (usually financial) to the affected community, to review value for money. For example, the costs of a livestock offtake intervention can be analysed against the cash received by livestock-selling households and the monetary value of any meat received.

LEGS recommends participatory approaches to evaluation, to ensure the engagement of affected communities in the description and analysis of response impacts. The participatory methods listed in *Appendix 3.1* can be adapted for impact evaluations as well as initial assessment and ongoing monitoring. See also *FAO* (2016): evaluation, pp. 170–174; impact evaluation, pp. 174–192; benefit-cost analysis, pp. 192–196.

6. Be accountable

MEAL systems should include mechanisms to enable feedback on the response from affected people. They should also ensure that the feedback is acted on and decisions communicated to those affected. The Core Humanitarian Standard (CHS) provides a framework for establishing and monitoring accountability based on nine commitments and associated quality criteria *(see Chapter 2: LEGS Principles, Table 2.1)*. Indicators to monitor performance against the nine commitments are also available. (See CHS and *References and further reading*.)

7. Share and apply learning

'Learning' in MEAL means that the information generated through monitoring and evaluation leads to analysis, reflection and changes in ongoing and future implementation. Learning ensures mistakes are not repeated, while lessons are used to adapt interventions to changing needs and contexts, as well as to influence future work.

Effective learning requires systems to be established to facilitate regular reviews of both formal monitoring data and informal experience. In addition to management, donor and government reporting, there should be regular community meetings. Otherwise, agencies should use other appropriate means to promote communication, participation and accountability with affected communities.

Key approaches include:

- creating a learning culture within the organisation and among staff – this includes committing time and resources to learning activities;
- establishing good communication processes that use appropriate languages and accessible formats;
- encouraging learning from previous emergency responses (for example, during the initial assessment);
- recording changes in implementation in response to feedback or operational context;
- sharing information with other implementers and coordination networks to increase learning.

(See CHS Commitment 7.)

Appendices

Appendix 3.1: Initial assessment – suggested participatory methods

Qualitative interview methods

Most participatory information collection techniques are based on qualitative interviews using a semi-structured checklist of open-ended questions, which facilitate discussion and follow-up. These interviews underpin the other techniques and generally take three forms:

- **Key informant interviews:** These target specific individuals representing particular groups or with particular knowledge, such as community leaders, local NGO and government staff, religious leaders, women's groups, and other civil society organisations.
- **Focus group discussions:** Groups of similar people (by gender, age, wealth, rank, interest group, livelihood strategy, etc.) are interviewed together, and then the same questions are repeated with similar and/or different groups to compare the findings.
- **Support for other participatory methods:** The discussion during participatory exercises such as mapping can provide useful additional insights; follow-up interviews may also be required to clarify the results of scoring or other exercises.

Visualisation methods

Carried out in focus groups, visualisation methods involve the use of local materials (either directly on the ground or on paper) to describe the local context and identify key issues:

- **Mapping** identifies key features of the area (for example, local resources, services, markets, grazing areas, water points, veterinary services), as well as other information such as insecure areas and livestock movements.
- **Seasonal calendars** show the timing of grazing migrations, linkages with cropping cycles, etc.
- **Historical timelines** identify significant events and crises (such as droughts or storms) that impact livelihoods.
- **Venn diagrams** illustrate the relationships between institutions or services.

LEGS framework
Emergency response planning

Ranking and scoring methods

Ranking and scoring methods provide numerical results that can be combined and analysed:

- **Proportional piling** shows relative values, for example of the impact of different livestock diseases.
- **Ranking** establishes an order of priority, for example of livestock problems.
- **Matrix scoring** is a way of marking against a range of criteria, for example scoring potential livestock interventions against speed of response, impact on livestock mortality, cost, etc.

Table 3.4 links some of the participatory methods outlined above with selected questions from the three initial assessment checklists.

Table 3.4: Suggested participatory methods for initial assessment

Assessment checklist*	Topic	Method
1.5	Gender/age roles and seasonality	Daily/seasonal calendar
2.5 2.6 2.8 2.9 2.10	Extent of affected area At-risk groups affected Services and facilities in both 'normal' times and in the emergency Natural-resource mapping (before and after): grazing; water; movements Impact on environment Seasonal changes	Mapping
2.3 2.4 2.7 2.8	Stages of the emergency Livestock disease trends Livestock sales trends Livestock price trends Livestock productivity trends	Timeline/time trend

continued over

LEGS framework
Emergency response planning

Assessment checklist*	Topic	Method
1.3 1.4 2.6 2.7 2.8	Sources of income/food Changes in nutritional status Changes in human disease Livestock sales, price, productivity changes	Proportional piling
1.3 1.4 2.7 3.1	Sources of income/food Livestock condition, morbidity, diseases History and effectiveness of previous response	Ranking/scoring
2.6	Affected population (to inform targeting)	Wealth ranking
1.6 3.2 3.3	Customary institutions' roles and relationships Key actors and coordination	Venn diagrams

* *Numbers refer to the questions from the three checklists*

Additional methodologies to support the initial assessment:

- **For question 1.5 on gender roles and power dynamics:** 'Women's empowerment in livestock index' (Galiè et al., 2018); see also *Box 1: Five Steps in a Rapid Gender Analysis* in the LEGS Gender Discussion Paper (de Jonge, K. and Maarse, L., 2020)

- **For question 2.6 on nutritional status:** *Standardized Monitoring and Assessment of Relief and Transitions Protocol* – an inter-agency initiative that provides reliable and consistent data on mortality, nutritional status, and food security. Consists of a survey manual and an analytical software program, supported by a database on complex emergencies – CE-DAT (SMART 2017)

- **For question 2.7 on livestock mortality:** see Catley et al. (2014) for an example of using *participatory epidemiology methods* to assess causes of livestock mortality during drought

**LEGS framework
Emergency response planning**

- **For question 2.9 on assessing the environmental impact of the emergency and any planned interventions:** See the *Rapid Environmental Assessment Tool* (Hauer, M. and Kelly, C 2018) and the FRAME assessment tool (UNHCR, 2009)
- **There are a number of resources for exploring coping strategies (questions 1.7, 2.6.5 and 2.6.6):** see, for example, the *Coping Strategies Index* (Maxwell and Caldwell 2008).

Further guidance on participatory methodologies is listed in *References and further reading*.

Appendix 3.2: Initial assessment – example checklist (partially completed)

Sample answers and data collection methods for selected questions from the initial assessment checklists are presented here for a rapid-onset emergency – a volcano in south-east Asia.

1.1 What are the main livelihood strategies in the affected area in 'normal' times?

The majority of households depend on livestock for their livelihoods, supplemented by crop production. The main livestock type is beef cattle (60 per cent of households in District A; 90 per cent of households in District B), while some also keep dairy cattle and pigs.

District	District A	District B
Human population	1,100,000	1,250,000
Total head of livestock	18,300	275,000

Methods: *provincial government reports (human and livestock populations); focus group discussion on livelihood strategies*

1.2 What are the key uses of livestock (for example, food, income, social, draught power, transport)?

- Meat is the most common use of livestock product, followed by milk in District B in particular.
- In both districts livestock are also a key source of draught power for crop production and manure for fuel and fertiliser.

Methods: *focus group discussion*

1.5 What roles do different household members play in livestock care and management (including use and disposal rights)? Pay particular attention to their gender and age. Take note too of different livestock species and ages as well as seasonal variations.

Men: responsible for management of draught animals for cultivation, care of the cattle.

Women: responsible for weeding and replanting the rice crop, care of poultry and small stock.

LEGS framework
Emergency response planning

Disposal rights are generally linked with these responsibilities. Women can sell small stock and their products, but disposal of large stock and the rice harvest is in the hands of the men.

In addition, previous eruptions from the volcano have caused sediment to flow into the nearby rivers, providing a natural resource for local communities. As a result, many of the men in the villages close to the rivers work as sand miners, while the women and other family members manage the livestock and crop production.

Methods: *daily and seasonal calendars; focus group discussion*

....

2.7 What has been the impact of the emergency on livestock? Differentiate by species if appropriate. Specifically:

2.7.1 How has livestock condition deteriorated?
Both districts were affected by ash and pyroclastic flows following the eruption, which affected the health and well-being of many livestock. All respondents reported the loss of some animals.

2.7.2 What is the impact on livestock welfare (for example, lack of feed and water; injuries; disease; extreme cold; or heat stress)?
In District A, the volcanic ash has significant health impacts on the livestock that survived, in particular diarrhoea (50 per cent of respondents) and respiratory problems (45 per cent). The shortage of available (or clean) livestock feed led to malnutrition in many livestock (30 per cent of respondents).

2.7.4 Has livestock morbidity increased?
The key causes of livestock death were:
– malnutrition due to limited access to feed and fodder (53 per cent);
– injury during evacuation (20 per cent);
– illness as a result of the ash and pyroclastic flows (15 per cent);
– dehydration (12 per cent).

Methods: *focus group discussion; proportional piling*

**LEGS framework
Emergency response planning**

Appendix 3.3: Initial assessment – recording template

The initial assessment findings can be recorded using a simple template, as in this example:

Date:

Location:

Participants – male:

Participants – female:

Name of interviewer:

Assessment method	Assessment question(s)	Key findings
Focus group discussion	1.1, 1.2	…..
Proportional piling	1.3	…..
…..		
…..		
…..		

LEGS framework
Emergency response planning

Appendix 3.4: PRIM – templates

Figure 3.3: Rapid-onset emergency PRIM template

Technical interventions	Scoring against LEGS livelihoods objectives			Appropriate timing for intervention		
	Immediate benefits	Protect assets	Rebuild assets	Immediate aftermath	Early recovery	Recovery
Feed						
Water						
Veterinary support						
Shelter						
Livestock offtake						
Provision of livestock						

3 – Very positive impact on objective
2 – Some impact on objective
1 – Very little impact on objective
0 – Not appropriate

Figure 3.4: Slow-onset emergency PRIM template

Technical interventions	Scoring against LEGS livelihoods objectives			Appropriate timing for intervention			
	Immediate benefits	Protect assets	Rebuild assets	Alert	Alarm	Emergency	Recovery
Feed							
Water							
Veterinary support							
Shelter							
Livestock offtake							
Provision of livestock							

3 – Very positive impact on objective
2 – Some impact on objective
1 – Very little impact on objective
0 – Not appropriate

For complex crises that include either a slow-onset or a rapid-onset emergency, the relevant PRIM may be used *(see PRIM Sample C in Appendix 3.5)*. For protracted or complex emergencies that do not include a slow-onset or rapid-onset crisis, only the left side of the PRIM (i.e., the livelihoods objectives) may be appropriate.

LEGS framework
Emergency response planning

Appendix 3.5: PRIM – samples

The following samples show how the PRIM can be used for different emergency types – rapid-onset, slow-onset and complex. In each sample, the PRIM matrix is followed by an explanation of the results.

Figure 3.5: PRIM sample A. Rapid-onset emergency – an earthquake in Asia

Technical interventions	Scoring against LEGS livelihoods objectives			Appropriate timing for intervention		
	Immediate benefits	Protect assets	Rebuild assets	Immediate aftermath	Early recovery	Recovery
Feed	1	3	3	✓	✓	—
Water	1	1	1	✓	—	—
Veterinary support	1	3	3	✓	✓	✓
Shelter	2	2	2	✓	✓	—
Livestock offtake	0	0	0	—	—	—
Provision of livestock	0	0	3	—	—	✓

3 – Very positive impact on objective
2 – Some impact on objective
1 – Very little impact on objective
0 – Not appropriate

Notes on PRIM Sample A:

- Providing feed immediately after the earthquake may contribute to household food security through milk or other livestock products from the surviving animals. Later, feed support may contribute to protecting and rebuilding livestock assets. If there is advance warning of the earthquake, some measures may be taken to stockpile feed and water.
- Providing water may offer some small benefit, depending on the effect of the earthquake on existing supplies.
- Veterinary support could provide immediate benefit, by helping to keep surviving animals alive in the immediate aftermath. It could also make a significant contribution to protecting and rebuilding livestock assets in the early recovery and recovery phases. It may help to maintain household/

maternal/child dietary quality where households rely on consumption of animal-source foods.

- Shelter-related interventions may contribute both to immediate benefits and to protecting and rebuilding assets, depending on the types of livestock kept and their shelter needs. If sufficient warning is given, moving them out of and away from buildings that may collapse may save livestock lives. In the immediate aftermath and early recovery phases, providing warm and/or dry shelter for affected animals is a significant contribution to the protection and rebuilding of assets.

- As the normal market system is not operating, commercial offtake cannot provide immediate benefits to crisis-affected households in this case. Slaughter offtake is most appropriate where livestock might otherwise die from lack of water or feed. It is therefore less likely to bring significant benefits to affected households in this instance. If animals are too emaciated for slaughter offtake to provide meat, they should be slaughtered on animal welfare grounds *(see Chapter 8: Livestock offtake)*.

- In terms of rebuilding assets, provision of livestock may contribute significantly by helping those who have lost their stock to begin to recover some livestock assets. However, this can only take place in the recovery phase.

LEGS framework
Emergency response planning

Figure 3.6: PRIM sample B. Slow-onset emergency – a drought in Africa

Technical interventions	Scoring against LEGS livelihoods objectives			Appropriate timing for intervention			
	Immediate benefits	Protect assets	Rebuild assets	Alert	Alarm	Emergency	Recovery
Feed	1	2	3	—	✓	✓	—
Water	1	2	1	—	✓	✓	—
Veterinary support	1	3	3	✓	✓	✓	✓
Shelter	0	0	0	—	—	—	—
Livestock offtake	3	2	1	✓	✓	(✓)	—
Provision of livestock	0	0	3	—	—	✓	✓

3 – Very positive impact on objective
2 – Some impact on objective
1 – Very little impact on objective
0 – Not appropriate

Notes on PRIM Sample B:

The slow-onset drought in Africa shows a very different pattern of interventions and timing compared with the Asian earthquake in PRIM Sample A.

- Providing feed and water during the alarm and emergency phases of a drought can help to protect remaining livestock assets and rebuild herds for the future.
- Animal health interventions, which may be carried out during all phases of a drought, can have a significant impact on protecting and rebuilding livestock assets. They do this by preventing death and disease and strengthening livestock resistance to drought.
- In this example, the provision of shelter is not appropriate.
- In the alert and alarm phases, commercial livestock offtake can contribute significantly to providing immediate benefits to affected families through the injection of cash. It can also contribute to a certain extent to protecting assets. This is because the remaining livestock have less competition for scarce resources, and some of the cash may be used to

support these remaining animals. If the timing of the intervention is delayed until the emergency phase, then commercial offtake may no longer be possible because the animals' condition will be too poor. In this case, slaughter offtake (shown by the tick/check mark in brackets) can provide some immediate benefits to affected households.

- In this example, because the drought is in the early stages (alert/alarm), the preference would be for commercial livestock offtake rather than slaughter offtake. This is because commercial offtake places cash in the hands of the livestock keepers, which encourages market processes and can help to protect other livelihood outcomes such as nutrition.
- In the recovery phase, providing livestock can make a significant contribution to rebuilding livestock assets.

The final sample PRIM shows how the combination of conflict and a slow-onset emergency can affect the appropriateness and feasibility of some of the potential interventions.

Figure 3.7: PRIM sample C. Complex emergency – drought with protracted conflict in Africa

Technical interventions	Scoring against LEGS livelihoods objectives			Appropriate timing for intervention			
	Immediate benefits	Protect assets	Rebuild assets	Alert	Alarm	Emergency	Recovery
Feed	1	3	3	—	✓	✓	—
Water	1	1	1	—	✓	✓	—
Veterinary support	1	3	3	✓	✓	✓	✓
Shelter	2	2	2	✓	✓	✓	✓
Livestock offtake	2	1	1	—	—	(✓)	—
Provision of livestock	0	0	3	—	—	✓	✓

3 – Very positive impact on objective
2 – Some impact on objective
1 – Very little impact on objective
0 – Not appropriate

LEGS framework
Emergency response planning

Notes on PRIM Sample C:

- Comparing this PRIM with Sample B, most of the possible interventions (such as feed, water, veterinary support, and provision of livestock) remain appropriate. Here they will also have the potential to deliver significant benefits to the affected communities.
- Providing feed has the potential to help protect and rebuild livestock assets. This is particularly true for communities confined to camps and unable to take their stock to pasture. Similarly, providing water for livestock that cannot be taken to the usual water sources because of insecurity may help to protect and rebuild livestock assets. This could also help protect household health and nutrition where water supply for human consumption is shielded from contamination by livestock.
- Shelter or enclosures for livestock, though irrelevant in PRIM Sample B, may become an important issue here because of displacement and insecurity (for example, the risk of looting).
- Commercial livestock offtake is not appropriate in this conflict situation since market systems and infrastructure are severely disrupted. Slaughter offtake (shown by the tick/check mark in brackets) could be possible, depending on the operational constraints under which agencies are working.
- All these interventions depend on the ability of responders to operate within the conflict situation.

Appendix 3.6: Cash and voucher assistance – response modalities, delivery mechanisms and decision tree

Figure 3.8: Response modalities for LEGS technical interventions and options

Technical interventions and options	Types of cash transfer		
	Cash grants*	Vouchers*	In-kind**
Feed			
Home-based emergency feeding	✓	✓	✓
Feed camp emergency feeding	✓	✓	✓
Water			
Water points	✓	✓	✓
Water trucking	✓	✓	✓
Veterinary support			
Clinical veterinary services	✓	✓	✓
Public sector veterinary functions	—	—	✓
Shelter			
Livestock shelter	✓	✓	✓
Livestock and settlement	—	—	✓
Livestock offtake			
Commercial livestock offtake	✓	—	✓
Slaughter livestock offtake for consumption	✓	—	✓
Provision of livestock			
Replacing lost livestock assets	✓	✓	✓
Building livestock assets	✓	✓	✓

*If implementing organisations are using cash grants or vouchers, they need to decide on whether to impose conditions (such as vaccinating livestock, attending training) or restrictions (such as specification of a type of feed or medicine) on the transfer so as to meet the objective of the programme. If cash transfers are large, such as in the provision of livestock, then a condition may be placed. For example, the total amount may be issued in tranches, with the release of each tranche dependent on evidence that livestock have been purchased with the previous one. Alternatively, if an objective has a specific element (such as livestock nutrition to maintain body condition), then agencies may impose a restriction to ensure livestock keepers buy appropriate feed of sufficient quality from specified vendors.

Cash grants may also be direct or indirect. A *direct* transfer may be made to livestock keepers, or an indirect transfer to, for example, a livestock trader. Commercial offtake may be considered as an *indirect* transfer, to support market processes, if agencies give loans to livestock traders. A *direct* transfer might be agencies providing vouchers to livestock keepers to purchase livestock.

** In-kind in this context may also include facilitation (for example, of livestock fairs), advocacy (on policy or licensing constraints), training, etc.

Table 3.5: Potential CVA delivery mechanisms for LEGS interventions

Category	Description	Possible provider
Cash grants		
Direct cash payment	Cash handed out directly to recipients by the implementing organisation	Implementing organisation/ partner
Delivery through an agent/ over-the-counter (OTC)	Cash delivered to recipients through a formal or informal institution that acts as an intermediary; does not require recipients to hold an account	Money transfer agents, post offices, traders, microfinance institutions, banks
Pre-paid card	Plastic card usable at cash machines (automated teller machines or ATMs), used for cash grants and vouchers; it can be swiped at point-of-sale devices, but it always requires network connection for transaction authentication	Banks, non-bank financial service providers, microfinance institutions, post office
Smart card	Plastic card with a chip, valid with point-of-sale devices and ATMs, used for cash grants and store purchases; can provide offline transaction authentication when network connectivity is off	Banks, non-bank financial service providers, post office

continued over

Category	Description	Possible provider
Mobile money	Encrypted code that can be cashed at various retail or other outlets, used for cash grants and vouchers; requires mobile network connection for transaction completion	Mobile network operator (MNO), banks
Bank account	Personal bank accounts or sub-bank accounts that are used to deposit cash grants; requires recipients to have formal identification (ID) documents and often formal residence status	Banks
Vouchers		
Voucher	Paper, electronic or some other form that can be exchanged for services or goods of a pre-determined value with pre-selected vendors	Implementing organisation/ partner/vendor

Adapted from/based on *Cash Delivery Mechanism Assessment Tool* – UNHCR
https://www.unhcr.org/5899ebec4.pdf

See CALP Network glossary in *References and further reading* for more information.

**LEGS framework
Emergency response planning**

Figure 3.9 Decision tree for CVA in livestock-based interventions

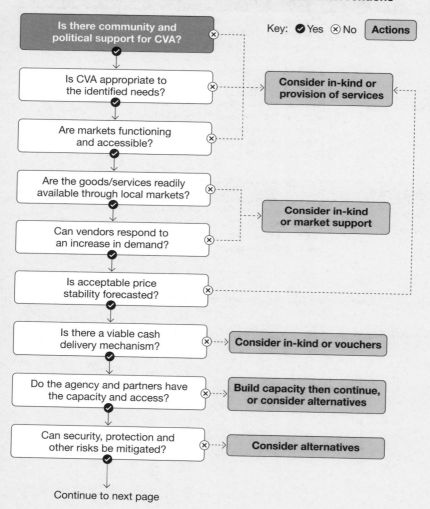

Continue to next page

LEGS framework
Emergency response planning

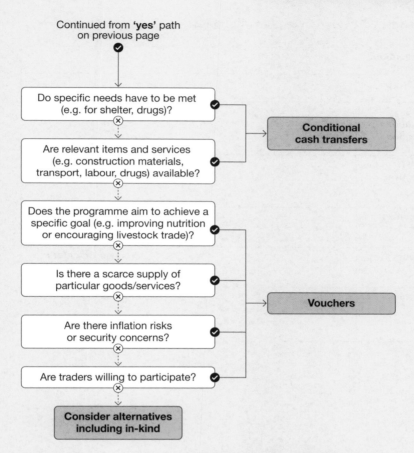

Decision tree adapted from *The Remote Cash Project Core Tool 2* (NRC, 2019) and the *ICRC Decision Tree for Cash Transfer Options* (found in Vetwork, 2011).

LEGS framework
Emergency response planning

Appendix 3.7: Response plan – template

Intervention:		
Option		
LEGS livelihoods objective		
SMART objective		**Impact indicators** • •
Outcomes		**Process indicators** •
Outputs		**Process indicators** •
Activities		**Process indicators** • •

Appendix 3.8: Response plan – example of livestock offtake intervention

Intervention: Livestock offtake			
Option	Commercial livestock offtake		
LEGS livelihoods objective	Protect livelihood assets		
SMART objective	At least 20% of income from livestock offtake is used to protect remaining livestock before the end of the intervention.		**Impact indicators** • Proportion of livestock offtake income used for water, feed, veterinary care or livestock transportation
Outcomes	Target households receive cash income from livestock offtake.		**Process indicator** • Average household income from offtake
Outputs	Livestock purchases		**Process indicators** • Average number and type of livestock purchased • Timing of purchases
Activities*	Identify target areas .... Identify livestock traders .... Facilitate community–trader dialogue .... Liaise with local authority on waiver of tax on livestock purchases	Targeting.... Timing... Locations... Procurement...	**Process indicators** • Number of meetings with community or with community representatives and other stakeholders • Number of community–trade meetings • Tax waiver in place

* The activities section of the response plan should contain as much detail as possible, including target groups, timings, locations, procurement processes and delivery modalities

**LEGS framework
Emergency response planning**

References and further reading

References and suggested resources for further reading are presented here, according to the following topics:

1. Assessment
2. Market analysis
3. Cash and voucher assistance
4. MEAL
5. Participatory methodologies

1. Assessment

ACAPS (2014) *Humanitarian Needs Assessment: The Good Enough Guide*, The Assessment Capacities Project (ACAPS), Emergency Capacity Building Project (ECB) and Practical Action Publishing. Rugby, https://www.acaps.org/humanitarian-needs-assessment-good-enough-guide

de Jonge, K. and Maarse, L. (2020) *Gender and livestock in emergencies:* a discussion paper for *the Livestock Emergency Guidelines and Standards* LEGS, https://www.livestock-emergency.net/wp-content/uploads/2020/11/LEGS-Discussion-Paper-Gender-and-Livestock.pdf

Food and Agriculture Organization of the United Nations (FAO) and International Labour Organization (ILO) (2009) *The Livelihood Assessment Toolkit: Analysing and Responding to the Impact of Disasters on the Livelihoods of People*, 1st edn, FAO and ILO, Rome and Geneva, https://www.fao.org/fileadmin/templates/tc/tce/pdf/LAT_Brochure_LoRes.pdf

Galiè, A., Teufel, N., Korir, L. et al. (2018) 'The women's empowerment in livestock index', *Social Indicators Research* 142: 799–825, https://doi.org/10.1007/s11205-018-1934-z

Hauer, M. and Kelly, C. (2018) *Guidelines for Rapid Environmental Impact Assessment in Disasters – Version 5, 2018* [Rapid Environmental Assessment Tool], Coordination of Assessments for Environment in Humanitarian Action Initiative, https://reliefweb.int/sites/reliefweb.int/files/resources/REA_2018_final.pdf

Maxwell, D. and Caldwell, R. (2008) *The Coping Strategies Index: Field Methods Manual,* 2nd edn, Cooperative for Assistance and Relief Everywhere, Inc. (CARE), https://www.researchgate.net/publication/259999318_The_Coping_Strategies_Index_Field_Methods_Manual_-_Second_Edition

SMART (2017) *SMART – Standardised Monitoring and Assessment for Relief and Transitions Manual 2.0*, SMART, Action Against Hunger Canada, and the Technical Advisory Group, https://smartmethodology.org/survey-planning-tools/smart-methodology/

Sphere (2014) *Sphere for assessments*, The Sphere Project and ACAPS, Geneva, https://spherestandards.org/resources/sphere-for-assessments/

2. Market analysis

Albu, M. (2010) *Emergency Market Mapping and Analysis Toolkit (EMMA)*, Practical Action Publishing, Rugby, https://www.emma-toolkit.org/toolkit

CALP (2018) *Minimum Standard for Market Analysis (MISMA)*, CALP Network, https://www.calpnetwork.org/publication/minimum-standard-for-market-analysis-misma/

International Red Cross and Red Crescent Movement (2014) *Rapid Assessment for Markets: Guidelines,* https://preparecenter.org/resource/rapid-assessment-for-markets-ifrc-guidelines/

SEEP (2017) *Minimum Economic Recovery Standards*, 3rd edn, the SEEP Network, Washington D.C., and Practical Action Publishing, Rugby, https://seepnetwork.org/Resource-Post/Minimum-Economic-Recovery-Standards-Third-Edition-exist-190

3. Cash and voucher assistance

The CALP Network website contains many tools and resources on cash and voucher assistance, including the Glossary of Terms and the Programme Quality Toolbox. See https://www.calpnetwork.org/resources/

Elluard, C. (2015) *Guidance notes: cash transfer in livelihood programming*, The CaLP Network and the Livelihoods Centre, https://www.calpnetwork.org/wp-content/uploads/2020/03/calp-waf-guidance-notes-en-web-1.pdf

FAO (2011) *The use of cash transfers in livestock emergencies and their incorporation into Livestock Emergency Guidelines and Standards (LEGS)*, Animal Production and Health Working Paper No. 1, FAO, Rome, https://www.fao.org/3/i2256e/i2256e00.pdf

Hermon-Duc, S. (2012) *MPESA project analysis: exploring the use of cash transfers using cell phones in pastoral areas*, Télécoms Sans Frontières and Vétérinaires Sans Frontières, Germany and Nairobi, https://www.calpnetwork.org/publication/mpesa-project-analysis-exploring-the-use-of-cash-transfers-using-cell-phones-in-pastoral-areas/

Norwegian Refugee Council (NRC) (2019) *Remote Cash Project Guidelines and Toolkit*, https://www.nrc.no/what-we-do/themes-in-the-field/cash-and-vouchers/remote-cash-project-guidelines-and-toolkit/

4. MEAL

Catley, A., Burns, J., Abebe, D. and Suji, O. (2014) *Participatory Impact Assessment: A Design Guide*, Feinstein International Center, Tufts University, Somerville, https://fic.tufts.edu/publication-item/participatory-impact-assessment-a-design-guide/

Cosgrave J., Buchanan-Smith M. and Warner, A. (2016) *Evaluation of Humanitarian Action (EHA) Guide*, ALNAP, https://www.alnap.org/help-library/evaluation-of-humanitarian-action-eha-guide

CHS Alliance, Group URD and the Sphere Project (2014) *Core Humanitarian Standard on Quality and Accountability*, 1st edn, https://corehumanitarianstandard.org/files/files/Core%20Humanitarian%20Standard%20-%20English.pdf

CHS Alliance, Group URD and the Sphere Project (2015) *CHS Guidance Notes and Indicators*, https://corehumanitarianstandard.org/files/files/CHS-Guidance-Notes-and-Indicators.pdf

Emergency Capacity Building Project (2007) *Impact Measurement and Accountability in Emergencies: The Good Enough Guide*, Oxfam, https://policy-practice.oxfam.org/resources/impact-measurement-and-accountability-in-emergencies-the-good-enough-guide-115510/

FAO (2016) *Livestock-related interventions during emergencies – The how-to-do-it manual*. Edited by Philippe Ankers, Suzan Bishop, Simon Mack and Klaas Dietze. FAO Animal Production and Health Manual No. 18. Rome, https://www.fao.org/3/i5904e/i5904e.pdf (See Chapter 10: Monitoring, evaluation and impact assessment)

LEGS (2016) *LEGS Evaluation Tool*, Livestock Emergency Guidelines and Standards, UK, https://www.livestock-emergency.net/legs-evaluation-tool/

LEGS framework
Emergency response planning

5. Participatory methodologies

Alders, R.G., Ali, S.N., Ameri, A.A., Bagnol, B., Cooper, T.L, Gozali, A., Hidayat, M.M., Rukambile, E., Wong, J.T. and Catley A. (2020) 'Participatory epidemiology: principles, practice, utility, and lessons learnt', *Frontiers in Veterinary Science* 7:532763, https://doi.org/10.3389/fvets.2020.532763

Holland, J. (ed.) (2013) *Who Counts? The Power of Participatory Statistics*, Practical Action Publishing, Rugby, https://practicalactionpublishing.com/book/2385/who-counts

LEGS (2021) *LEGS Participatory Techniques Toolkit*, Livestock Emergency Guidelines and Standards (LEGS), https://www.livestock-emergency.net/wp-content/uploads/2022/02/LEGS-Participatory-Techniques-Toolkit_May-2021.pdf

Pretty, J.N., Guijt, I., Thompson, J. and Scoones, I. (1995) *Participatory Learning and Action: A Trainer's Guide*, International Institute for Environment and Development (IIED), London, https://www.iied.org/6021iied

UNHCR and CARE (2009) *Framework for Assessing, Monitoring and Evaluating the Environment in Refugee-Related Operations* (FRAME Toolkit), joint UNHCR and CARE project, https://www.unhcr.org/uk/protection/environment/4a97d1039/frame-toolkit-framework-assessing-monitoring-evaluating-environment-refugee.html

See also case studies for planning livestock-based interventions in emergencies at: https://www.livestock-emergency.net/resources/case-studies/

Technical standards

Livestock feed standards
1. Preparedness
2. Assessment and planning
3. Feeding levels
4. Feed safety

Water standards
1. Preparedness
2. Assessment and planning
3. Location of water points
4. Rehabilitation and establishment of water points
5. Sources and quality for water trucking
6. Logistics and distribution for water trucking

Veterinary support standards
1. Preparedness
2. Assessment and planning
3. Clinical veterinary service design
4. Examination and treatment
5. Zoonotic diseases
6. Sanitation and food hygiene
7. Disposal of dead animals
8. Livestock disease surveillance

Shelter and settlement standards
1. Preparedness
2. Assessment and planning
3. Livestock shelter
4. Livestock and settlement
5. Transition to durable livestock shelter solutions

Livestock offtake standards
1. Preparedness
2. Assessment and planning
3. Commercial livestock offtake
4. Slaughter livestock offtake for consumption

Provision of livestock standards
1. Assessment and planning
2. Defining the intervention package
3. Delivery systems
4. Additional support to recipients

Technical standards
Livestock feed

Chapter 4: Technical standards for livestock feed

132 **Introduction**

135 **Options for ensuring feed supplies**

141 **Timing of interventions**

142 **Links to other LEGS chapters and other HSP standards**

143 **LEGS Principles and other issues to consider**

148 **Decision tree for livestock feed options**

151 **The standards**

161 **Appendix 4.1: Assessment checklist for feed provision**

163 **Appendix 4.2: Examples of monitoring and evaluation indicators for livestock feed interventions**

165 **References and further reading**

Photo © Michael Benanav

Technical standards
Livestock feed

Chapter 4: Technical standards for livestock feed

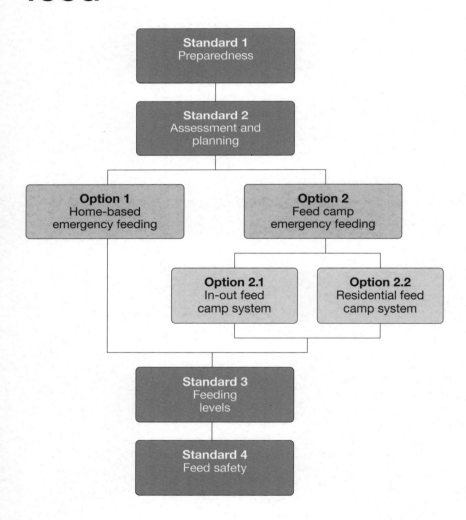

Technical standards
Livestock feed

Introduction

Different types of emergencies, and their severity, affect access to feed resources for livestock. For example, drought emergencies reduce soil moisture levels, suppress plant growth, and reduce the amount of pasture and forage available for livestock. In contrast, volcanic ash and flood or storm water can cover and kill pasture and forage resources. Even when abundant, access to pasture and forage resources may be lost during times of conflict. Some emergencies that affect access to livestock feed may last just a few weeks, but others may extend over years.

When an emergency reduces the availability of feed, the livestock that are affected need to be provided with adequate, timely and quality supplementary feed. They are then protected from weight loss and can continue to contribute to household food production. In pastoral settings, continued access to milk can have very positive outcomes on child nutrition and health.

This chapter presents information on the importance of livestock feed as an emergency livestock response. It also provides technical options for feed interventions, along with the associated benefits and challenges of each. LEGS does not specifically address fodder production but acknowledges the importance of feed systems, including fodder production during non-emergency times. Information is also available in Chapter 6, 'Provision of Feed', in *FAO* (2016). For each technical option LEGS provides information through standards, key actions and guidance notes. Checklists for assessment, as well as monitoring and evaluation indicators, are presented as appendices at the end of this chapter. A list of further reading is also provided. Case studies are presented on the LEGS website (see https://www.livestock-emergency.net/resources/case-studies/).

Links to the LEGS livelihoods objectives

Ensuring livestock have access to adequate feed during emergencies supports each of the three LEGS livelihoods objectives:

- Feeding productive animals may contribute to immediate improvements in household food supply, thus **supporting crisis-affected communities to obtain immediate benefits using existing livestock assets**.
- Feeding livestock, including draught and other working animals, may result in the **protection of key livestock assets**.
- Feeding core breeding animals supports the rebuilding of herds and flocks, thus **rebuilding key livestock assets**.

Technical standards
Livestock feed

The importance of ensuring feed supplies in emergency response

As with human populations, livestock depend on feed and water. Productive animals are particularly vulnerable to disruptions in feed and water supplies (see *Chapter 5: Water*). Any intervention that aims to protect livestock affected by an emergency must, therefore, consider access to key feed resources, at least to limit livestock losses. The provision of livestock feed is particularly important where emergencies result in reduced access to forage resources. In such situations, excess livestock deaths are the result of starvation, as opposed to livestock disease. This is important in drought-affected areas, but also in the case of volcanic eruption, extensive flooding and conflict.

Access to supplementary livestock feed may also be lost during emergencies, when supply lines are disrupted or feed stores destroyed. This may result from a cyclone, earthquake or flood, or through conflict. Following such emergencies, it may be necessary, as part of the provision of livestock feed, to re-establish supply lines and rebuild feed stores and markets to maintain livestock productivity in the longer term. (See *Process case study: Animal feed banks in Niger for drought preparedness.* This case study includes investment in establishing livestock feed services that can continue after the emergency.)

In times of emergency, some humanitarians regard the delivery of feed to animals as having a negative effect on saving human lives. This is because the delivery of food aid and other essential non-food items to human populations may be disrupted, especially if transport resources are limited. The importance of livestock feed is, however, recognised by livestock keepers. It is not uncommon for livestock keepers around the world to share food aid with their productive animals. Others may sell a portion of their food aid in the local markets to purchase feed to protect key livestock assets.

There are cost implications associated with providing emergency livestock feed over an extended period, such as through a multi-year drought. Costs therefore need to be assessed carefully before a feed intervention is started. Cost concerns should be set against alternatives, however, perhaps using a simple benefit-cost analysis tool. This will help confirm the appropriateness of a feed intervention compared to alternatives, such as the long-term provision of food aid or restocking. Costs of nutritional interventions to support pastoral children who no longer have access to milk, or other humanitarian assistance, can be compared against the cost of providing replacement animals.

For example, a benefit-cost study was carried out in the pastoral areas of south-eastern Ethiopia and north-east Kenya following drought after two failed rains. The study found it was between three and six times more expensive to restock a household with a core herd of small ruminants than to protect the same number of animals by providing livestock feed.

The timely provision of livestock feed during emergencies also contributes to the first of the animal welfare five domains, namely 'nutrition' – factors that involve the animal's access to sufficient, balanced, varied and clean food and water. This is described in *Chapter 1: Introduction to LEGS*. (See also the *Impact case study: Measuring the impacts of cattle supplementary feeding in Ethiopia*. This case study confirms significantly reduced mortality rates among cattle that received supplementary feed.)

The well-timed provision of emergency livestock feed to protect core breeding animals, when forage and other feed resources are lost or inadequate, can strengthen measures already initiated by livestock keepers themselves. In times of drought, for example, livestock may be grazed on roadside verges. (Verges benefit from rainfall run-off from the road; this then increases moisture availability and supports better pasture production.) Similarly, during times of flood, livestock may be grazed on roadside embankments (that are raised and therefore remain flood free). In times of emergency, livestock may also be moved to mountains and riverine areas where better pasture is available. To supplement these resources, livestock keepers may also purchase agricultural by-products from local food processing plants to feed to their livestock.

Despite such measures, it may not be possible for livestock keepers themselves to compensate fully for feed lost in an emergency. In such cases, they may need additional livestock feed. This can include conserved forage such as hay, crop residues, concentrate feeds, multi-nutrient blocks (MNB), or home-produced mixed feeds such as cactus. In some areas, home-produced feeds have the added benefit that they can use invasive plants that reduce pasture availability. (See *Process case study: Using invasive plants for animal feed in Sudan*.) In addition, assistance may be required for the rebuilding of damaged or destroyed livestock feed stores. This is particularly important for areas affected by volcanic activity, where livestock feed may be contaminated by falling ash. It will also be important for markets, especially in those areas where markets play a central role in the provision of livestock feed.

Technical standards
Livestock feed

Options for ensuring feed supplies

When planning an emergency livestock feed intervention, it is important to consider transport costs. For loosely bundled crop residues, such as sorghum, millet or sugar cane tops, or if straw or hay bales are used, the transport costs are high. This is particularly so when these feeds are transported over long distances, and used to feed large animals such as cattle, camels and equines (which consume a relatively large amount of feed each day). When the quality of forage is poor, with low nutritional value, the impact of the intervention will also be low.

Costs may be inflated when feed suppliers learn that feed resources are to be delivered under an emergency response intervention. To help ensure better value for money, it may be helpful to use higher nutrient value concentrate feeds, home-produced mixed feeds, or multi-nutrient blocks. MNBs – and pelleted concentrate feeds – also have an advantage over concentrates in powder form. This is because they are less likely to be blown away or lost to livestock if fed on dusty ground.

The type and level of assistance provided will typically vary depending on the role that livestock play in local livelihoods of those affected by an emergency. For example, an agency may be able to provide a modest number of smallholder farmers, or households living in peri-urban areas, with adequate feed resources to meet all their livestock feed needs. This is because they may own one or two milk cows, or a pair of plough oxen, a working animal, or fewer than 10 sheep and goats. In this way, smallholder farmers and peri-urban milk producers/transporters using working animals may be able to continue in their previous livelihoods.

In contrast, it is less likely that an agency or group of agencies will able to provide adequate feed resources to meet the needs of a large pastoral community affected by a multi-year drought. This is because the combination of the number of livestock and duration of the feeding interval is simply beyond agencies' financial and logistical capacities. In such cases, pastoralists may target the survival of a smaller number of core breeding animals and therefore ensure that some animals are safeguarded for the future.

Where it is necessary to target only a small number of core breeding animals, it may be helpful to link a feed response with a livestock offtake intervention. This removes surplus animals (for which there are no emergency feed resources available). At the same time, commercial offtake animals may help generate cash and will reduce operational costs. Similarly, slaughter offtake for consumption may provide an important source of meat for households dependent on food aid. In this way, the costs of providing a nutritionally balanced diet are reduced, and the costs of the slaughter offtake for consumption can be spread *(see Chapter 8: Livestock offtake)*. The removal of livestock (through either form of livestock offtake) will help ensure that feed resources are not diluted across a higher number of livestock than planned.

Irrespective of careful planning, sound logistics, and linking to livestock offtake, emergency livestock feed interventions are costly. They are also logistically and managerially challenging. It is therefore important to agree and plan intervention objectives with emergency-affected livestock keepers. The intervention objectives may be to reduce livestock body weight loss, maintain body weight, recover lost body weight, maintain production levels or increase production levels.

It is also important to plan exit strategies from the outset. By doing so it may be possible to control potentially spiralling costs and avoid the associated early and untimely closure of emergency livestock feed interventions.

Most documented feed interventions are for ruminants; a summary of the associated benefits and challenges is presented in *Table 4.1*.

Table 4.1: General benefits and challenges of feed interventions

Benefits	Challenges
Key livestock assets are protected	Sourcing feed during an emergency may prove challenging
The need for long-distance migration that may trigger conflict is reduced	Purchase and transport costs of feed may be inflated in times of emergency
Protected livestock may supply milk/eggs or create income through ploughing/transport during an emergency	The supply of livestock feed through an emergency may be expensive and unsustainable
Protecting livestock assets may support flock/herd rebuilding and protect livelihoods	The routine supply of emergency feed may artificially increase local livestock populations
Access to family livestock following an emergency may help keep families together, benefit well-being and reduce trauma	Increased livestock populations can place greater demands on scarce water resources

continued over

Technical standards
Livestock feed

Benefits	Challenges
	Access to safe storage and reliable transport systems may be limited as a result of the emergency
	The transport of different feed resources may result in the spread of different plant pests and diseases.
	The importation of feed from outside the area may disrupt local feed markets; for example, causing shortages and higher prices in areas where procurement takes place

There are two main technical options associated with the emergency provision of livestock feed: 'home-based' and 'feed camp'. For both options, agreement needs to be reached first on which animal types and the number of participating animals. The livestock keepers (both men and women) will then want to decide for themselves which individual animals to protect as their core breeding animals for the future. (See *Process case study: Women help manage a nucleus herd feeding programme in Ethiopia.* This case study illustrates women being paid a small salary to work in the residential feed camps. In this way, they are encouraged to participate in the emergency response.)

Option 1: Home-based emergency feeding

Under this technical option, feed resources are purchased, transported, stored and distributed to participating households. The feed comes from a central distribution point such as the local agricultural office, agricultural cooperative, or market centre. Alternatively, feed can be distributed to individual homes through assisted and subsidised secondary forms of transport, such as small trucks, tractors/trailers, or working animals.

In addition, agencies can support home-based emergency feeding interventions through cash and voucher assistance (CVA). Livestock keepers (both men and women) are provided with either cash or vouchers to purchase feed for their animals. Such schemes are feasible only when markets are functioning or can be supported to function. The safe movement of community members and transport of purchased feed should also be ensured. Cash grants may be provided unconditionally, although the intended use for the cash is communicated to livestock keepers. This allows for different types of feed to be bought depending on the type of livestock. Alternatively, the provision of a commodity voucher would place restrictions on where the feed can be bought and the type of feed to be purchased.

A summary of the benefits and challenges associated with home-based emergency livestock feed interventions is presented in *Table 4.2*.

Option 2: Feed camp emergency feeding

Under this technical option, emergency feed is provided to selected livestock in one of two feed camp sub-options:

Sub-option 1: *The 'in-out' feed camp system* to which livestock keepers bring an agreed type and number of livestock (for example, two lactating cows, five sheep/goats or one camel). Animals may be brought to the feed camp daily or every other day, according to the planned level of feeding and intervention outcomes *(see Standard 3: Feeding levels)*.

Typically, selected animals are marked (often with coloured and numbered ear tags) to ensure that the same animals routinely participate. Livestock keepers are discouraged from rotating different animals into the intervention. This is because rotating animals dilutes the effectiveness of emergency feeding and may reduce survival rates.

Sub-option 2: *The 'residential' feed camp system* to which an agreed type and number of livestock are brought and remain within a camp setting. When the emergency is over, animals are returned to their keepers and again become their primary responsibility.

Again, animals entering the camp are typically marked so that they are not confused with other animals when taken to water (if water is not available on site).

Table 4.2 presents a summary of the general benefits and challenges associated with feed camp livestock feeding, and the specific benefits and challenges associated with the different sub-options.

Technical standards
Livestock feed

Table 4.2: Benefits and challenges of feed supply options

Benefits	Challenges
Home-based emergency feeding	
Livestock keepers retain management of their livestock	Feed losses can occur at each transport/handling stage
Livestock can forage locally in the daytime and receive daily rations in the evenings	There are additional costs and logistical challenges associated with home delivery (albeit borne by the affected communities)
Home-produced milk/eggs or ploughing/transport benefits the family	Affected communities may dilute feed resources – giving it to additional animals and neighbours
Complex camp management and associated logistical challenges are avoided	The provision of veterinary care is more challenging
	Monitoring the use of feed and associated improvements in animal body condition is more difficult
Feed camp emergency feeding	
Livestock keepers are required to select core animals for protection	There are increased requirements (compared with 'home-based' option) – animal health specialists and services, troughs/feeders, access to adequate water, fences, protected feed stores
Measured rations ensure proper use of available feed	
Security is in place for livestock and community members	Security provision is required
Improved veterinary care can be provided	Increased costs of key support staff – feed camp manager, other technical and liaison staff, and security personnel
There are income-generating opportunities for caretakers, guards	Management challenges occur – including managing expectations of local elite
Easier for monitoring use of feed and improvements in animal body condition	Compensation payments are typically paid for livestock deaths in camps

continued over

Benefits	Challenges
Sub-option 1: The in-out feed camp system	
There are reduced night-time security costs	Livestock need to be healthy to be able to trek to the feed camp daily
Keepers are responsible for watering livestock	There are additional herding demands for selected animals to be brought to and from the camp
Feeding can be staggered to avoid congestion and minimise disease transmission	
Sub-option 2: The residential feed camp system	
There is limited/no dilution of feed resources by livestock keepers	Requires organised labour to fence, feed, water and secure livestock
Animals are isolated and therefore disease risks are reduced	Daily organised labour is required for the duration of the emergency

Timing of interventions

Taking into account the costs involved, emergency livestock feed interventions result in higher benefit-cost ratios when they are short-term in nature. They can, for example, provide important 'bridging' interventions after floods that can be expected to subside within a short period. Similarly, feed interventions can support drought-affected livestock until the next rains, when soil moisture levels are replenished, and natural forage again becomes available.

It may become necessary to provide feed resources for an extended period, perhaps even a year or more. This will happen in more severe droughts when, for example, even the next seasonal rains are forecast to be poor, or during protracted and complex emergencies. Typically, the costs of such interventions are prohibitive for all but a small number of core breeding animals. While such interventions may help protect local breeds, they will do little to help protect pastoral livelihoods.

Technical standards
Livestock feed

The possible timings of emergency feed interventions for different emergencies are presented in *Table 4.3*. In non-emergency phases, agencies and livestock keepers may invest in forage and feed production systems and the utilisation of feed. This process should build on locally proven good practice, including rangeland management. The *Process case study: Regenerating pasture as a preparedness activity to protect core breeding animals, Ethiopia* describes progress made through enclosure of degraded rangeland areas, the construction of water-harvesting structures, and seeding with local forage species. Harvested and dried grass was subsequently shared among community members.

Table 4.3: Possible timing of livestock feed interventions

Options	Rapid-onset emergency		
	Immediately after	Early recovery	Recovery
1. Home-based	✓	✓	—
2. Feed camp	✓	✓	—

Options	Slow-onset emergency			
	Alert	Alarm	Emergency	Recovery
1. Home-based	—	✓	✓	—
2. Feed camp	—	✓	✓	—

Links to other LEGS chapters and other HSP standards

Reference has been made to the important links between providing emergency livestock feed and livestock offtake *(see Chapter 8: Livestock offtake*; see also *Process case study: Complementary feed provision and livestock offtake in Niger)*. Through support for livestock offtake, surplus animals are removed from the production system. Providing livestock feed can then ensure the survival of a small number of core breeding animals.

Retaining people's livestock assets is fundamental to restoring livelihoods, contributing to longer-term food security and to improved nutrition. This is because it increases their ability to manage other potential causes of undernutrition. Protected core breeding animals also require animal healthcare *(see Chapter 6: Veterinary support)* and water supplies *(see Chapter 5: Water)*. To ensure different interventions complement each other, it is important that all initiatives are coordinated.

LEGS Principles and other issues to consider

Table 4.4: Relevance of the LEGS Principles to livestock feed interventions

LEGS Principle	Examples of how the principles are relevant in livestock feed interventions
1. Supporting livelihoods-based programming	The provision of feed protects livestock-dependent livelihoods from being lost when feed resources are provided to at least the level for maintaining livestock body weight. Feed targeted at key livestock assets, including core breeding animals, secures livelihoods for the long term. Where feed interventions allow livestock to remain productive, livestock actively supply nutrition for households and wider communities. Feed interventions support additional livelihoods including those involved in the feed supply line – transporters, suppliers of supplementary feeds/concentrates, and anyone involved in feed camps (caretakers, guards, etc.).
2. Ensuring community participation	Community participation is essential for the selection of participating households and the appropriate livestock for feed interventions. Ongoing active community participation helps ensure the appropriate and equitable use of the feed resources provided. Community involvement in feed camp management is also critical for their success.

continued over

Technical standards
Livestock feed

LEGS Principle	Examples of how the principles are relevant in livestock feed interventions
3. Responding to climate change and protecting the environment	Feed interventions should prioritise support for local breeds.
	Feeding should be staggered and aligned with provision of water to avoid high concentrations of livestock. It should be done in conjunction with the use of 'home-based' feed options where possible.
	Environmental impacts need to be considered during the establishment of feed camps.
	It is important to avoid spreading plant pests and diseases from other areas during the transportation of feed supplies.
	Smallholder farmers may benefit from organic matter provided by livestock, either in feed camps or when they are brought to graze crop residues.
4. Supporting preparedness and early action	The timely procurement, transport and pre-positioning of feed resources is essential for effective interventions.
	Support for haymaking and irrigated fodder production can strengthen resilience during non-emergency phases and increase preparedness. Communities can also benefit from early warning systems and fully functional contingency funds that can be activated when required.
	Early interventions reduce livestock stress and support animal welfare.
5. Ensuring coordinated responses	All interventions should be appropriately coordinated to ensure efficient use is made of all available resources, as well as integration with broader humanitarian assistance.
	Complementary and coordinated interventions, including livestock offtake, veterinary support, and water, increase the effectiveness of all technical interventions.

continued over

LEGS Principle	Examples of how the principles are relevant in livestock feed interventions
6. Supporting gender-sensitive programming	Women, men and youth should participate fully in each stage of the intervention: design, implementation and management, as well as monitoring, evaluation, accountability and learning (MEAL). This will avoid gender-bias in the targeting of appropriate livestock types and the numbers of participating households. Special attention should be given to women-/child-headed households, who may be overlooked.
7. Supporting local ownership	Feed interventions should identify potential for localising the intervention as much as possible. This includes local production, local transporters and warehouse operators, and also local staffing and management of feed camps.
8. Committing to MEAL	Routine monitoring of the provision of livestock feed is necessary to ensure cost-effective and quality supply, effective distribution systems, and adequate rations/nutrition levels for selected livestock. Benefit-cost analysis of the provision of livestock feed versus other emergency responses is important. Evaluations to promote accountability and the sharing of positive and negative results from feed interventions are also essential.

Targeting at-risk groups

Research in southern Africa confirms that during drought emergencies, poorer livestock keepers lose more of their livestock (as a proportion of their herds) than richer livestock keepers. They also spend more per animal on average to protect them. It is therefore important for interventions to ensure that poorer livestock keepers (with fewer livestock per household) receive an adequate share of livestock feed resources. This can be done through the appropriate design and management of emergency feed interventions.

Where such controls are inadequate, feed resources are typically diverted by the wealthy and influential, who secure disproportionate amounts for their own animals. While no less at risk, such households have access to greater resources to protect and provide for their livestock. At another level, inadequate controls may encourage non-livestock keepers to divert whole shipments of livestock feed for sale to other livestock keepers.

Technical standards
Livestock feed

Supporting local capacities and coping strategies

Livestock-keeping communities affected by emergencies invariably draw on their knowledge, skills and capacities to respond. In anticipation of seasonal floods, smallholder farmers may move their livestock from flood plains to higher ground. Additional livestock may be saved when timely and clear official alerts are issued, with smallholder farmers moving their animals to safety in anticipation. Where such alerts are not given, large releases of water from dams can result not only in the loss of field crops but also of livestock, and damage to livestock-related infrastructure too.

Similarly, during the onset of drought, pastoralists may trek their livestock to protected drought reserves. When these fail, they may trek again to areas less affected by drought. To support such long-distance treks, pastoralists tap into extended social networks. Increasingly too, they have to negotiate with local and national governments. The African Union and sub-African regional economic groups are progressively recognising the huge economic benefits reaped from supporting livestock mobility. In West Africa, for example, cross-border mobility rights are enshrined in law through pastoral codes and the demarcation of livestock migratory routes. Support for mobility, including customary cross-border mobility, helps strengthen local coping strategies – as many pastoral areas of the world are located near international borders. However, this kind of support needs to consider security risks for livestock keepers, relations with host communities (including competition for available livestock feed), and increased threats from the spread of livestock diseases. Hence, external support to customary relocation strategies requires thorough assessment and analysis before it can be adopted as an intervention.

Immediately after an emergency, it is not uncommon for livestock keepers to sell older and very young animals *(see Chapter 8: Livestock offtake)*. They may also prevent remaining animals from breeding, to protect females from the nutritional demands made by unborn young.

Recognising the importance of local capacities and coping strategies, it is important for external agencies to follow local guidance and to support and consolidate local knowledge, skills and capacities *(see LEGS Principle 7: Local ownership)*.

Further specific examples of livestock keepers' knowledge and skills are highlighted in the key actions and guidance notes below.

Minimising the introduction of pests and diseases

During an emergency livestock feed intervention, it may be necessary to import feed from another area (though preferably not from another country). During such interventions, seeds of invasive plants, pests and diseases may be transported and introduced. To minimise risk, it is important to ensure independent quality assurance and high-quality phytosanitary management at the point of purchase and dispatch.

Considering disruption of local markets

While recognising the benefits of emergency livestock feeding, it is important the intervention does not interrupt local livestock feed systems and markets. Interventions should give priority to sourcing livestock feed locally and supporting local suppliers (for example in the district/region). In this way, local feed systems and markets can continue to support livestock production after the emergency feed intervention is ended (see *Process case study: Animal feed banks in Niger for drought preparedness*, described above).

A market assessment will help determine the best option and approach. Such an assessment would clarify the amount of feed that can be purchased and over what time interval without negatively affecting prices and supplies. In some emergencies, it may be necessary to develop a hybrid model to ensure continuity of supply, purchasing some feed resources locally, while importing others.

By using cash- and voucher-based approaches, agencies can help support and strengthen local livestock feed markets. As already indicated, this will also help support continued improvements to livestock production after the emergency has ended.

Camps

Displaced livestock keepers may arrive in Internally Displaced Persons (IDPs) or refugee camps with some of their livestock. Once resident, they may also acquire new animals. IDPs and refugees living with livestock require additional support so they can feed and water their animals *(see Chapter 5: Water)*. They will also need essential animal healthcare *(see Chapter 6: Veterinary support)*.

The host community may also include livestock keepers. In this situation, the arrival of large numbers of additional livestock into the area is likely to put further strain on available livestock feed and water resources. This may result in competition that can erupt into conflict. To help ensure good relations with

**Technical standards
Livestock feed**

the displaced community, it may be helpful to also provide the host community with some supplementary livestock feed or animal healthcare.

Within the camp too, men and women with livestock require additional space to ensure that the immediate area does not become soiled *(see Chapter 7: Shelter)*. In some cases, it may be helpful to relocate households with livestock to areas with more space and access to pasture and feed resources. In this way, possible tensions with the host community and with other displaced communities can be avoided.

The arrival of additional livestock in the area is not entirely negative. In some areas, smallholder farmers may welcome the opportunity to have livestock grazing crop residues post-harvest, as this improves soil fertility through the addition of organic matter.

Decision tree for livestock feed options

The decision tree *(Figure 4.1)* summarises some of the key questions to consider in determining which may be the most feasible and appropriate option for an emergency livestock feed intervention. The standards, key actions and guidance notes that follow provide more information for detailed planning. Where possible, they build on preparedness activities conducted prior to the onset of the emergency, or in 'normal' times.

Technical standards
Livestock feed

Figure 4.1: Decision tree for livestock feed options

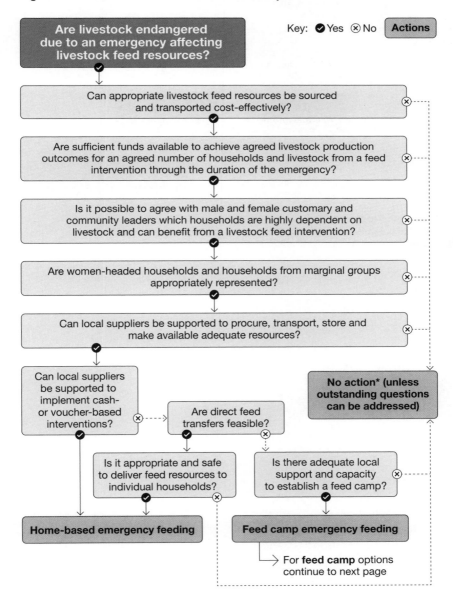

**Technical standards
Livestock feed**

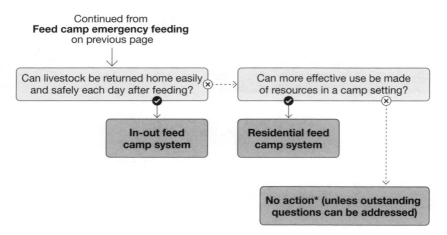

*The result **'No action'** does not necessarily mean that no intervention should take place, but rather that further training or capacity strengthening may be required in order to be able to answer 'yes' to the key questions.

Technical standards
Livestock feed

The standards

Standard 1: Preparedness

Preparedness activities are undertaken in advance of the emergency to ensure the timely procurement, transport and pre-positioning of feed resources.

Key actions

- Undertake participatory assessments that consider local livestock owner knowledge and skills, and local adaptive and coping strategies *(see Guidance note 1)*.
- Conduct a thorough pre-intervention availability assessment to identify potential local sources of feed and analyse the level of risk of market disruption *(see Guidance note 2)*.
- Where feeds will need to be sourced externally and stockpiled, identify pre-approved suppliers with reliable and sustainable sources, giving due consideration to quality and price *(see Guidance note 3)*.
- Conduct proper security assessments along proposed feed supply routes and the areas of distribution *(see Guidance note 4)*.
- Identify implementing agencies that have adaptable administrative systems and procurement processes to allow them to purchase feed efficiently and effectively *(see Guidance note 5)*.
- Identify locally recognised community and/or administration structures that will be able to ensure fair and transparent distribution approaches *(see Guidance note 6)*.
- Consider the wider policy constraints that might affect access to feed in emergencies, and advocate for customary strategies *(see Guidance note 7)*.

Guidance notes

1. Local coping strategies

For pastoralists, mobility is the primary livelihood strategy for accessing widely dispersed seasonal grazing and water. In times of drought, mobility is typically extended further and involves the trekking of livestock to remote grazing areas. It may even involve moving across international borders into neighbouring countries. In some cases, pastoralists transport livestock long distances in lorries. There is little evidence to demonstrate the impact of

Technical standards
Livestock feed

agency support for enhancing migration to improve livestock survival rates. However, the protection of migratory routes will almost certainly help improve the overall resilience of pastoral production systems. Emergency feed response plans can enable local livestock keepers to gain recognition and provide support for their well-established adaptive and coping strategies, such as mobility.

Most smallholder farmers collect and store crop residues during the harvest season. These can be fed to livestock during the next dry/cold season, when naturally occurring livestock feed is limited. During emergencies, they may not be able to collect and store crop residues, or the amounts they have will be inadequate. Support for livestock offtake and provision of feed will help smallholder farmers balance animal numbers and availability of livestock feed.

2. Availability of local feeds

The pre-identification of locally available livestock feed offers significant advantages over imported feeds:

- Cost: Locally available feeds are typically less expensive (although purchase costs may be higher in the local area immediately affected by the emergency).
- Transport costs: Shorter distances result in reduced transport costs.
- Transport losses: Shorter distances reduce the likelihood of theft and damage/wastage during transportation where feed is loaded and unloaded on multiple occasions.
- Disruptions that result from imported feeds 'seeping' into local markets, are avoided.
- Cash may be injected into the local economy through feed purchases.
- Local labour: There may be increased opportunities for the use of local labour in the harvest, transport, storage and distribution of locally procured livestock feed.
- Local capacity: Support for local feed producers and markets can help drive innovation.

On the other hand, local procurement may also result in competition between agencies, local feed suppliers and local livestock keepers for available feed resources. Typically, such competition results in quality compromise, inflated feed prices and distorted markets that in turn draw imports into the area.

3. Sourcing feeds externally and stockpiling

Some emergency feeding programmes may benefit from the use of concentrate feeds with specific nutritional formulations or MNBs, which cannot be sourced locally. In rare cases, these may have to be sourced from cities or even, in some cases, neighbouring countries as a short-term measure. To ensure the safe delivery of high-value feeds, appropriate transport and infrastructure must be available. To further minimise potential risks, interventions should include the following:

- arranging adequate in-country storage facilities and pre-positioning of feed to allow stockpiling and cover for interruptions to deliveries (stockpiling is, however, not without risks of pilfering as well as feed contamination or degradation; effective feed store management is key);
- identifying and using more than one supply chain so that the failure of one does not completely disrupt the intervention;
- assessing the availability of local alternatives for short-term use as a stopgap; for example, locally sourced high-protein cottonseed or oilseed cakes, mixed with crop by-products and naturally available cactus and *Prosopis* pods, might be an effective substitute for costly imported concentrates *(see Process case study: Using invasive plants for animal feed in Sudan)*;
- adopting more modest nutritional and production objectives for an emergency feeding programme that might be satisfied using locally available mixed feed.

4. Security

Minimising security risks to community members and staff in the procurement, storage, transport and distribution of livestock feed takes precedence in the delivery of all emergency livestock feed interventions (see *Chapter 2: LEGS Principles* on the Sphere Protection Principles). It should therefore be part of preparedness planning. Most international agencies have well-established security guidelines, and many collaborate with security companies to ensure the well-being of their staff. In contrast, local agencies with more limited resources find it difficult to achieve a similar level of protection. In complex emergencies, international organisations may partner with these smaller, less well-resourced agencies that are on the front line. Agencies may reduce risks by contracting local transporters and pre-approving them to deliver to the final point of distribution. The reason for the reduced risk is that such contractors typically have strong social networks.

Technical standards
Livestock feed

5. Administrative systems

Agencies with restrictive procurement systems that do not easily allow feed purchase from small-scale suppliers may need to upgrade their administrative systems. It may be possible, for example, under an agency's emergency preparedness planning process, to develop a preferred suppliers list and to support due diligence checks.

6. Distribution structures

Where possible, the distribution of daily or weekly feed rations should be managed by men and women representing local community structures and institutions. Customary elders (including men and women) should help recruit reliable local staff and assist agencies to target participating households.

7. Policy/advocacy context

Preparedness activities should include a thorough policy assessment that takes account of possible obstacles to the purchase, transport and delivery of livestock feed, particularly in emergency contexts (for example, restrictions on moving livestock feed across national and district boundaries). Where local coping strategies exist but are not being used, an analysis of hindering forces can also help identify the constraints and prevent an inappropriate intervention. Advocacy may also help reinstate such strategies. For example, it may be possible in the non-emergency phase of drought and flood emergencies for interventions to help establish drought/flood reserves as part of a wider preparedness plan. In this way, a feed intervention can help address chronic livestock feed shortages, support unintended positive outcomes and gain increased levels of local engagement and ownership *(see Process case study: Animal feed banks in Niger for drought preparedness)*.

> ### Standard 2: Assessment and planning
> Assessment of options for the provision of emergency livestock feed is informed by the role livestock play in local livelihoods; livestock feed needs; and the opportunities for delivering an appropriate intervention.

Key actions

- Initiate emergency livestock feed interventions only where there is a high probability that affected communities will continue to keep and benefit from livestock after the emergency has ended *(see Guidance note 1)*.

Technical standards
Livestock feed

- Select key livestock assets for feeding based on an analysis of their importance in local livelihoods, their health status, the chance of their surviving the emergency and their longer-term usefulness in rebuilding livestock assets in the future *(see Guidance note 2)*.
- Where it is necessary to establish feed camps, ensure appropriate security, ease of access (for livestock and livestock keepers), and storage facilities for feed and livestock medicines. Ensure too that logistics and resources are sufficient to support the camp for the duration of the emergency. Management of both livestock and feed resources should also support the earliest possible return to normal livestock-keeping practices *(see Guidance note 3)*.
- Explore the use of CVA during the initial assessment *(see Guidance note 4)*.
- Consider linking with other emergency interventions such as child nutrition, livestock offtake, provision of water for livestock, and veterinary support *(see Guidance note 5)*.

Guidance notes

1. Affected communities continue to keep livestock

Some men and women affected by the emergency may be at risk of losing all their livestock assets. This may be either through the direct loss of animals or a reduction in household capacity through injury, ill health, migration or death. Or they may no longer be able to keep (or be interested in keeping) livestock. Therefore, before launching an emergency livestock feed intervention, it is important to be sure that participating households want to continue to keep livestock. Community-led household targeting will identify those most likely to benefit in the medium and longer term from the emergency livestock feed intervention. Others who will not benefit should be provided with alternative assistance.

2. Targeting livestock

Some animal types are better adapted to coping with and recovering from feed and water shortages than others. Camelids and goats, for example, are far better at surviving drought conditions than cattle and sheep. It may therefore be strategic for interventions to target camelids and goats, and in this way support the transition to more drought-resistant animal types. Reference has already been made to the need to target core breeding animals. It may also be appropriate for interventions to target working animals that continue to carry water or supplies (including fodder) during emergencies. Once the animal type and number have been agreed, however, livestock keepers themselves should be supported to select the individual animals that will participate in the intervention.

3. Feed camps

Establishing feed camps requires significant investment (both financial and in-kind contributions). So it is important that local institutions and potential participants are fully involved in all design and planning phases. Issues to be considered include whether available resources are adequate for the anticipated duration of the emergency feeding intervention, security, accessibility and exposure/shelter.

Management issues similarly should be addressed in an inclusive and participatory manner. These issues include construction of camp infrastructure (feed and water troughs/racks); handling systems for sick animals; the management of feed stores; the daily handling of feed and water; the provision of animal health services; and staff accountability procedures. Where at all possible, overall managerial responsibility for feed camps should be under local community control, with external technical support as needed (see *Process case study: Women help manage a nucleus herd feeding programme in Ethiopia*).

4. Use of cash and vouchers

CVA schemes work well for households with limited storage capacity. However, they should be living in areas where livestock feed markets are working well, and where quality feed resources are available. Vouchers are particularly useful in ensuring regular access to fresh feed. While appropriate, there are potential challenges. For example, traders may find themselves on the receiving end of inflated prices as producers seek to inflate profits. Careful monitoring of feed availability, quality and cost is therefore required to ensure that cash and voucher mechanisms can continue and also ensure value for money. For local traders and suppliers involved in voucher schemes, an additional small financial incentive may be appreciated for their administrative time.

5. Complementary programming

Where appropriate, the provision of livestock feed may be integrated into a broader programme of humanitarian assistance that includes child nutrition, livestock offtake, and the provision of water and veterinary support.

Appendix 4.1: Assessment checklist for feed provision contains a checklist to guide the assessment and planning process.

Technical standards
Livestock feed

Standard 3: Feeding levels

Levels of emergency feeding should ensure livestock survival, meet agreed production outcomes, and be sustainable over the life of the intervention.

Key actions

- Determine feeding levels for the intervention with reference to planned nutritional aims for different livestock types *(see Guidance notes 1 and 2)*.
- Ensure feeding levels by the intervention are both attainable and sustainable *(see Guidance note 2)*.
- Where the loss of feed markets and stores represents an immediate or short-term threat to future livestock production, replenish reserves as part of the emergency livestock feed intervention *(see Guidance note 3)*.

Guidance notes

1. Nutritional adequacy

This concept recognises the difference between a maintenance diet – which will keep an animal alive – and a production diet. The latter will support growth, sustain a pregnancy, or support milk/egg production or draught animal work. A maintenance diet for a small ruminant, however, will be very different from a maintenance diet for a cow or a camel. For details of different definitions, feed components and suggested diets (including for equids, poultry and pigs) see *FAO* (2016), Chapter 6.

The higher the livestock production level, the more costly the intervention. Hence, the higher the production level, the smaller the number of animals it will be possible to feed and protect, and the shorter the duration of the response. Given a choice, most livestock keepers will likely choose to protect more animals and therefore agree to a reduced-value diet. However, if there are nutritional benefits for children from continued milk production, it may be possible to justify increased spending and support for a production diet.

2. Feed budgeting

Planning the quantities of feed involves balancing the daily feed requirements of the participating animals, the purchase and transport costs, the duration of the planned feed intervention, and the intervention budget, as follows:

- daily feed requirements – the daily feed needs of different animal types, based on the planned nutritional aims of the intervention as described under *Guidance note 1*;

- available feed resources – the quantities of different feeds available locally or transported from other areas;
- purchase price – the cost per unit;
- transport cost – the distance from the source of feed to the final point of delivery (the central store or distribution points, in other words), and the cost per unit;
- provision of feeding bowls and measuring tins – to ensure the best use is made of available feed resources;
- duration – the duration of the proposed programme;
- budget – the financial resources available for the intervention;
- people/staff required to deliver the intervention.

Based on the above information, it is possible to determine the number of animals that can be supported through the intervention. Where resources are limited, it may be necessary to reduce the number of participating animals, the duration of feeding or the nutritional aims. The alternative is to seek additional funding.

3. Feed store replenishment

Rapid-onset or complex emergencies may destroy livestock feed stores and disrupt feed markets and supply lines. Where these losses immediately threaten livestock survival or continued production, interventions can include the reconstruction of stores and the replenishment of supply lines. This ensures the protection of key livestock assets, and their continued production.

> **Standard 4: Feed safety**
>
> Where feed is purchased from outside the affected area, proper sanitary, phytosanitary, and other aspects of feed safety ensure the delivery of quality feed resources.

Key actions

- Assess risk levels in local livestock populations, and naturally occurring pasture and forage sources, to imported pests, diseases, and vectors (see *Guidance note 1*).
- Screen feed materials at the point of purchase for significant pests, diseases, and other sources of contamination (see *Guidance note 2*).

- Implement appropriate measures to ensure that vehicles and storage facilities are clean and sanitary *(see Guidance note 3)*.

Guidance notes

1. Risk assessments

It may be difficult to undertake a risk assessment during an emergency. Nevertheless, this is both necessary and important before committing to an emergency livestock feed intervention. Where the risks associated with a particular feed source are deemed high, it may be better to redesign the intervention and source feed resources elsewhere.

It is important to document practices and lessons of emergency feed interventions, to ensure that lessons are learned (see *Chapter 3: Emergency response planning*, section on MEAL).

2. Quality control

Wherever it is sourced, all emergency livestock feed must be quality assured. This includes visual checks for weed seeds, pests, dust, sand, soil particles, and metals; and laboratory checks for microscopic fungal growth, including aflatoxin contamination. Feed samples tested in a laboratory should also be tested for dry matter, crude fibre, mineral matter, crude protein, crude fat, and calcium content (particularly important for lactating animals).

3. Sanitary procedures

It is generally better to ensure that quality control measures are carried out at the point of purchase and dispatch rather than at the point of delivery. As a minimum, quality control measures should include the following:

- Quality assurance: Receive and file quality assurance papers from the supplier, and ensure feed products are appropriately bagged and labelled before transport.
- Washing and airing: Between loads, all lorries/trucks and storage facilities need to be washed and aired.
- Record keeping: To avoid cross-contamination, keep records of materials carried by lorries/trucks. For example, livestock feed should never be transported in trucks that have previously carried hazardous materials such as agrochemicals, glass, lubricants or scrap metal.
- Minimal contact: Keep human contact with emergency livestock feed to a minimum. Community members should not, for example, be transported on lorries that are also transporting livestock feed.

Technical standards
Livestock feed

- Protection: When being transported, all livestock feed should be covered by waterproof tarpaulins or covers to avoid dust and rain.
- Timelines: Transport and storage times should be kept to a minimum.

Appendices

Appendix 4.1: Assessment checklist for feed provision

This checklist can be used during rapid initial assessments to help inform the appropriateness and design of emergency livestock feed interventions.

For all emergency feed interventions

Rations and nutritional quality

- Have daily feed rations been developed that are appropriate to the specific production objectives of the feed intervention? See *FAO (2016)*, p. 95.
- Does the planned feed regime take full account of the logistical difficulties that may be encountered when attempting to deliver to target communities?
- Do these feed rations take realistic account of available budgets?

Feed safety

- Are feed quality assurance and standard operating procedures in place?
- Have feed quality assessments been conducted, including for possible weed seed, pests, diseases and other feed contaminants that increase risks associated with the intervention?
- Are the quality control measures (for example, routinely taking samples to a laboratory for analysis) for screening feeds adequate?
- Are storage times for feeds consistent with maintaining feed safety, quality, and warehouse management?
- Are proper procedures in place for ensuring adequate standards of cleanliness both for vehicles used for transporting feeds and for storage facilities?

Sourcing and distribution of feeds

- Are the agencies' administrative systems flexible enough to meet the needs of a continuing feed supply intervention?
- Where possible, has feed been sourced locally to minimise production and transport costs, and to support local producers, traders and other businesses?
- Where feeds are sourced locally, have steps been taken to ensure that other stakeholder groups are not put at risk as a result?

**Technical standards
Livestock feed**

- Has provision been made for the replenishment of depleted feed stores during the recovery phase?
- Can opportunities for backloading (for example, livestock offtake) be identified to ensure trucks carry loads both into and out of affected areas, so increasing the efficiency of the distribution system?
- Are transport routes, warehouse and distribution networks adequately protected from security risks?

For emergency feed camps

Acceptability of feed camp and identification of affected communities

- Are the local community, their representatives and local administration actively participating in the intervention and regularly consulted?
- Have participating households (both male- and female-headed) been fully and appropriately informed of what the planned feed camp can – and cannot – offer, and the terms under which they would participate?
- Have potential participating men and women been properly informed about the risks to which they might be exposed because of participating in the initiative?
- Are potential participating households likely to be able to meet the demands of participation (such as providing labour for overseeing animals)?
- Are proper procedures in place for identifying participating households and the most appropriate animal types to be targeted by the establishment of a feed camp?

Logistics and management

- Can construction and other materials necessary for establishing the feed camp be sourced locally or transported to the site at an acceptable cost, time and with minimal risk?
- Are adequate supplies of feed and water available or deliverable for the level of occupancy envisaged for the camp?
- Can appropriate support services be provided, such as animal health?
- Are there managers with appropriate levels of skills and commitment available to run the camp?
- Are management structures in place that can address the needs and concerns of all local stakeholders?
- Can adequate levels of staffing be put in place for the camp? (Where possible, labour inputs should include participating communities.)

Appendix 4.2: Examples of monitoring and evaluation indicators for livestock feed interventions

	Process indicators (measure things happening)	Impact indicators (measure the result of things happening)
Designing the intervention	Number of meetings with livestock keepers, community representatives and other stakeholders, including, where relevant, private sector suppliers	Roles and responsibilities of different actors
		Approach for feed provision, including procurement, transport, and distribution
		Community engagement in selecting participating households and number and type of livestock to receive feed
		Community involvement in managing livestock to receive feed (e.g. in village-based feeding centres)
Implementation	Amount and value of feed procured and delivered to feed sites	Comparative body condition scores of animals receiving versus not receiving feed
	Number and type of livestock receiving feed	Mortality rates in animals receiving versus not receiving feed
	Amount of feed by type of animal per day	Changes in child nutrition
	Duration of feeding	Changes in women's/girls' time requirements in collecting feed
	Tendering process – procurement and transport	Evidence of influence on humanitarian policy
	Administrative system that supports local purchases	Reference to feed quality, quality assurance and compliance with quality phytosanitary standards – point of purchase, transport, storage, and delivery to end point/user
	Availability assessment	Routine use of laboratory analysis
	Security assessment	Quality assessments
	Cash and vouchers	

continued over

Technical standards
Livestock feed

	Process indicators (measure things happening)	Impact indicators (measure the result of things happening)
Implementation *(continued)*		Local purchases
		Security incidents
		Local administrative systems/institutions
		Inclusivity of local structures
		Cash/vouchers

References and further reading

Ayantunde, A.A., Fernández-Rivera, S., and McCrabb, G. (eds) (2005) *Coping with Feed Scarcity in Smallholder Livestock Systems in Developing Countries*, International Livestock Research Institute (ILRI), Nairobi, http://mahider.ilri.org/handle/10568/855

Bekele, G. and Abera, T. (2008) *Livelihoods-based Drought Response in Ethiopia: Impact Assessment of Livestock Feed Supplementation*, Feinstein International Center, Tufts University, Digital Collections and Archives, Medford, MA, http://hdl.handle.net/10427/71144

Bekele, G., Ali, T. E., Regassa, G. and Buono, N. (2014) 'Impact of goat feeding and animal healthcare on child milk access in Ethiopia', *Field Exchange* 47, p72, https://www.ennonline.net/fex/47/goat

Catley, A. (ed.) (2007) *Impact Assessment of Livelihoods-based Drought Interventions in Moyale and Dire Woredas, Ethiopia*, Pastoralists Livelihoods Initiative, Feinstein International Center, Tufts University, Medford, MA, together with CARE, Save the Children USA, and USAID-Ethiopia, https://fic.tufts.edu/assets/IMPACT1-2.pdf

Catley, A., Admassu, B., Bekele, G. and Abebe, D. (2014) 'Livestock mortality in pastoralist herds in Ethiopia and implications for drought response', *Disasters* 38(3): 500–516 https://doi.org/10.1111/disa.12060

Cullis, A. (2021) *Understanding the Livestock Economy in South Sudan: field study findings*, TANA Copenhagen, https://tanacopenhagen.com/wp-content/uploads/2021/04/Field-Report.pdf

FAO (2021) *Pastoralist Knowledge Hub* https://www.fao.org/pastoralist-knowledge-hub/en/

FAO (2016) *Livestock-related interventions during emergencies – The how-to-do-it manual.* Edited by Philippe Ankers, Suzan Bishop, Simon Mack and Klaas Dietze. FAO Animal Production and Health Manual No. 18. Rome, https://www.fao.org/3/i5904e/i5904e.pdf (See Chapter 6)

Gebru, G., Yousif, H., Adam, A., Negesse, B. and Young, H. (2013) *Livestock, Livelihoods, and Disaster Response, Part Two: Three Case Studies of Livestock Emergency Programmes in Sudan, and Lessons Learned*, Feinstein International Center, Tufts University, Medford, MA, https://fic.tufts.edu/publication-item/livestock-livelihoods-and-disaster-response-part-two-2/

Goe, M.R. (2001a) *Assessment of the scope of earthquake damage to the livestock sector in Gujarat State, India*, Consultancy Mission Report, FAO, Bangkok/Rome.

Goe, M.R. (2001b) *Relief and rehabilitation activities for the livestock sector in earthquake affected areas of Kachchh District, Gujarat State, India*, Technical Cooperation Project Proposal, FAO, Rome/Bangkok.

International Institute for Environment and Development (IIED) and SOS Sahel UK (2009) *Modern and Mobile: The Future of Livestock Production in Africa's Drylands,* de Jode, H (ed), IIED and SOS Sahel UK, London, https://pubs.iied.org/12565iied

Krätli, S. (2015) *Valuing Variability: New Perspectives on Climate Resilient Drylands Development*, de Jode, H (ed), IIED, https://pubs.iied.org/sites/default/files/pdfs/migrate/10128IIED.pdf

LEGS (2018) *Revisiting the economic impacts of early drought response: How does early response affect households in pastoralist area?* LEGS Briefing Paper, prepared by A. Catley, LEGS https://www.livestock-emergency.net/general-resources-legs-specific/

Ngaka, M.J., 2012, 'Drought preparedness, impact and response: a case of the Eastern Cape and Free State provinces of South Africa', *Jàmbá: Journal of Disaster Risk Studies* 4(1), Art. 47, https://jamba.org.za/index.php/jamba/article/view/47

Technical standards
Livestock feed

PASTRES: Pastoralism, Uncertainty, Resilience [website], https://pastres.org

Sadler, K., Mitchard, E., Abdulahi, A., Shiferaw, Y., Bekele, G. and Catley, A. (2012) *Milk Matters: The Impact of Dry Season Livestock Support on Milk Supply and Child Nutrition in Somali Region, Ethiopia*, Feinstein International Center, Tufts University and Save the Children, Addis Ababa, http://fic.tufts.edu/publication-item/milk-matters/

Samanta, A.K., Ali, Y., Jahan, F.N. and Hossain, B. (eds) (2021) *Feeding and Health Care for Livestock During Natural Calamities*, South Asia Association for Regional Cooperation, Dhaka, Bangladesh, https://www.researchgate.net/profile/Kuenga-Namgay/publication/351476488_Feeding_and_healthcare_of_livestock_during_natural_calamities_in_Bhutan/links/6099f743299bf1ad8d90ac0a/Feeding-and-healthcare-of-livestock-during-natural-calamities-in-Bhutan.pdf

See also case studies for feed interventions in emergencies at: https://www.livestock-emergency.net/resources/case-studies/

Technical standards
Livestock feed

4

**Technical standards
Provision of water**

Chapter 5: Technical standards for the provision of water

- 170 **Introduction**
- 173 **Options for water provision**
- 177 **Timing of interventions**
- 178 **Links to other LEGS chapters and other HSP standards**
- 179 **LEGS Principles and other issues to consider**
- 182 **Decision tree for water provision options**
- 186 **The standards**

- 199 **Appendix 5.1: Assessment checklist for water points**
- 201 **Appendix 5.2: Examples of monitoring and evaluation indicators for water provision**
- 202 **Appendix 5.3: Considerations for water point management**
- 203 **References and further reading**

Photo © Kelley Lynch

Chapter 5: Technical standards for the provision of water

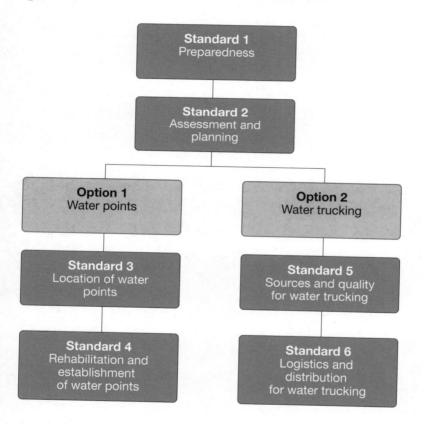

Introduction

Water, alongside feed and veterinary services, is an essential element that keeps animals alive within all types of husbandry systems. Productive animals are extremely vulnerable to disruptions in their water supply. In the absence of water, most animals (except for some camelids) cannot survive for more than a few days. When emergencies result in the loss of natural water sources, or physical damage to artificial water sources, access to water becomes a critical issue for livestock-keeping communities.

Emergency responses that support the provision of water keep livestock alive. They also ensure livestock productivity is retained for crisis-affected households (for example, through the supply of milk and eggs). The technical options available for water provision vary depending on whether the emergency is slow-onset or rapid-onset. Pre-emergency preparedness ensures that assistance or support that targets existing water supplies, or focuses on trucking in water from outside areas, is undertaken effectively. Preparedness also means that support will not create further challenges for affected communities.

This chapter presents information on the importance of providing water as a livestock emergency response. It also presents the technical options for water interventions and the associated benefits and challenges of each. Information is also available in Chapter 7, 'Provision of water', in *FAO* (2016). For each technical option, LEGS provides information through standards, key actions and guidance notes. Checklists for assessment, as well as monitoring and evaluation indicators, are presented in the appendices at the end of this chapter. A list of further reading is also provided. Case studies are presented on the LEGS website (see https://www.livestock-emergency. net/case-studies).

Links to the LEGS livelihoods objectives

The provision of water for livestock in an emergency ensures the survival of livestock assets during and beyond the emergency. It relates to all three of the LEGS livelihoods objectives:

- Healthy livestock contribute to household food security. So the provision of water relates to the first LEGS livelihoods objective by **supporting crisis-affected communities to obtain immediate benefits using existing livestock resources.**
- As with the provision of feed, livestock vital to livelihoods are kept alive by the provision of water during emergencies. This therefore relates to the second LEGS objective, **to protect key livestock assets.**
- In addition, the intervention supports existing and new livestock assets through preventive, resilience-strengthening measures. These include the timely rehabilitation of water points and water sources to maximise capacity for storage and access. This is therefore in line with the third LEGS livelihoods objective, **to rebuild key livestock assets.**

Technical standards
Provision of water

The importance of the provision of water for livestock in emergency response

During emergencies, where water sources have been seriously compromised, the provision of water is an extremely important intervention. It is also critical to get it right, as poorly considered responses can lead to environmental degradation and the exacerbation of conflict.

In many communities, access to water for livestock is often extremely variable: a reflection of the pre-existing climatic conditions as well as inadequate public supplies. In the semi-arid and arid environments that support low/highland pastoralism or mixed farming systems across Africa and Asia, water supply is highly seasonal. Drought is a major factor, causing a slow-onset emergency that frequently combines with conflict to create a complex emergency. Tropical and subtropical climates have relatively high rainfall and more reliable year-round water supply. Here, rapid-onset emergencies (such as earthquakes, but also floods) are more likely to cause loss of water access. These losses occur through damage to water supply facilities as well as water pollution through sewage, rubbish, silt, mud and carcasses. (Earthquakes in more arid areas, including Pakistan, Iran and Afghanistan, can also become complex emergencies when combined with conflict.)

In many dry areas, livestock water points (including boreholes, wells and dams) are now increasingly unable to support levels of demand. Water point degradation results from chronic decreases in the water table, lack of maintenance or spare parts, intentional destruction, failed management, and higher concentrations of animals. Long-term development to reduce fragile and intermittent water access is often a complicated and highly contested issue, linked to regional and national economic, energy and agricultural policies. Within agriculture, the long-term water policy trend is towards commercialisation, privatisation and large-scale irrigation, creating further challenges for communities.

When responding to critical shortages in livestock water supply as part of an emergency intervention, implementing agencies need to be aware of the causes and broader issues in water access. They must also recognise the very limited window for action if livestock are to survive. Preparedness is crucial.

While water for livestock must meet some basic quality requirements, the quality standard is not as high as it is for human consumption. Livestock can make use of water that is unfit for humans. For specific animal daily water quantity and quality requirements, see *FAO* (2016, p. 113).

The provision of water also contributes to the first of the five animal welfare domains, namely, 'nutrition' – factors that involve the animal's access to sufficient, balanced, varied and clean food and water. This is described in *Chapter 1: Introduction to LEGS*.

Options for water provision

As a general rule interventions should be based on the most cost-effective and sustainable option for water provision. Rehabilitating or establishing new **water points** often offers the most viable, long-term solution to water shortages, assuming that their use is in line with LEGS *Principle 3: Environmental protection*. However, when the need to deliver water is acute, an intervention may have to include expensive and unsustainable methods such as **water trucking**, at least for the short term. Despite the initial high costs, providing trucked water is more economically viable than replacing lost livestock assets, and may be the more feasible option.

Delivery mechanisms such as vouchers, mobile money or cash payments – conditional or unconditional, restricted or unrestricted – may be appropriate and cost-effective for water provision. Depending on the market and availability, livestock keepers may use them to purchase private sector, community-owned or other livestock water supplies. (See *Chapter 3: Emergency response planning* for details on cash and voucher assistance – CVA.) However, direct cost recovery from affected communities should also be considered. Humanitarian organisations now increasingly encourage affected communities to make payments directly so that existing payment systems for water are not undermined. This direct payment option should be explored for both water points and water provision via trucking. Doing so is more likely to lead to a more sustainable delivery chain than externally provided funding.

Option 1: Water points

Water points for livestock take several different forms, including wells, boreholes, and surface water harvesting systems (for example, check dams and storage tanks). Urban mains supplies may also be used.

Technical standards
Provision of water

During an emergency, the intervention may secure access to water points for livestock keepers in one of three ways:

1. changing the management of existing water points to provide broader access to crisis-affected communities;
2. rehabilitating existing but degraded water points;
3. establishing new water points.

Each of these approaches can present difficulties. While the first approach could be implemented for the lowest cost, it may not be feasible due to insufficient water supply. Or it may not be feasible because of the complexities of meeting the needs of existing users as well as new users from crisis-affected communities. In slow-onset emergencies, rehabilitation of degraded water points may be appropriate as a preparedness intervention. In rapid-onset emergencies, there may not be sufficient time for rehabilitation, unless the work required for improving water availability and/or quality is relatively minor and spare parts are readily available. Establishing new water points can be complex and time-consuming. Here, issues including site location, social agreements, excavation, management, and confirmation of community ownership make it more feasible as a long-term development activity.

Conflict between the water demands of human populations and their associated livestock may also be an issue in all three approaches. This also applies to existing or new conflicts between those with crop-based livelihoods (see SEADS) and those with livestock-based livelihoods. However, with proper planning and management it should be possible to create a network of distribution points that can meet the needs of both humans and animals.

Option 2: Water trucking

Water trucking should generally be regarded as a last-resort option and only used during the first stages of an emergency. It is expensive, resource-inefficient and labour-intensive. Due to the critical impact of dehydration on livestock, however, it is sometimes the only option that can be implemented rapidly to keep animals alive in the short term. As a rule, therefore, trucking should be regarded as a temporary intervention to be replaced as soon as possible by other means of water provision. Opportunities for cost sharing/cost recovery should also be explored.

Wherever possible, after thorough analysis and assessment, livestock owners should contribute in cash or 'in kind' to the costs of water trucking. This can lead to **partially subsidised** water trucking, with communities managing contributions and organising the trucking themselves. However, where communities are unable to contribute, agencies may need to **fully subsidise** water trucking. A sound communication strategy is essential for this approach, since all stakeholders need transparent and equal information flows.

Water trucking involves major logistical inputs, and great care and attention need to be given to the planning and management of trucking operations. This includes monitoring the evolving situation, making sure that routes remain open, that drivers and other crew are protected from changes in the security situation, and that tankers are maintained effectively.

More details on the benefits and challenges of the different options for the provision of water are shown in *Table 5.1*.

Table 5.1: Benefits and challenges of water provision options

Option	Benefits	Challenges
1. Water points		
1.1 Changing management of existing water sources	This is a relatively cheap option, making maximum use of existing opportunities and resources	There are often limited opportunities on the ground to achieve this
	It can normally be implemented rapidly in response to an emergency	It has the potential to introduce conflict among groups of new users
1.2 Rehabilitating existing water sources	This is potentially cheaper than other water provision options	The reasons for original degradation (e.g. low water table) may still apply or recur
	Management structures and systems for the water source may already exist	It takes time, and sourcing materials for rehabilitation works will often take too long
	It is a long-term solution that can outlast the emergency	

continued over

Technical standards
Provision of water

Option	Benefits	Challenges
1.2 Rehabilitating existing water sources *(continued)*	It has potential to provide water for both livestock and human needs It provides support to at-risk households through cash-for-work initiatives (e.g. dam desilting, cleaning natural water catchments, rehabilitation of existing pan)	
1.3 Establishing new water sources	This has the potential to provide sustainable new water sources for emergency and post-emergency populations in immediate locality of need It has the potential to provide water for both livestock and human needs	It is more costly than rehabilitation, requiring very high capital investment and long-term financing The time needed is likely to be too long for an effective response to the emergency Appropriate siting may be difficult in a short (emergency) time frame Locally based and agreed management systems need to be established to prevent conflict and ensure equitable access, and to ensure sustainable use of the water resource and the surrounding environment There are potential negative consequences (conflict, environmental degradation) of making new areas accessible to people and livestock There are risks due to modification of the usual grazing pattern (easy access to dry-season pastures, modification of migration routes, land tenure disputes, etc.)

continued over

Technical standards
Provision of water

Option	Benefits	Challenges
2. Water trucking		
	This can respond rapidly to immediate water needs	It is expensive and resource-inefficient; moving livestock to water sources where there is still enough animal feed may be more appropriate
	It may make use of water of insufficient quality for human consumption	It is labour-intensive and logistically complex
		It is not sustainable – temporary solution only
		It offers the greatest potential for conflict between human and livestock water needs
		It requires a locally based management structure to ensure equitable access to water
		There is potential for conflict with existing users of water source

Timing of interventions

Water trucking is a short-term measure that may be appropriate in the immediate aftermath of a rapid-onset emergency, or in the emergency phase of a slow-onset emergency. It should not be continued beyond these stages, as it is a costly and unsustainable intervention. Changing water point management or the rehabilitation of existing points, in contrast, may be carried out in all stages of both emergency types. The establishment of new water sources is a feasible solution when existing degraded water sources are insufficient or unsuitable for rehabilitation. This intervention should ideally link with longer-term water development programmes and improved management, as part of both emergency preparedness and post-emergency response. *Table 5.2* suggests appropriate timing for each of these water options.

Resilience-strengthening measures play a significant role in humanitarian responses and should be considered during stable non-emergency periods.

~~ Technical standards
Provision of water

Preparedness contributions (for example, tools, spare parts) provided during 'normal' times can help alleviate an acute phase of water shortages during an emergency. Other preparedness activities are suggested in *Standard 1: Preparedness* below.

Table 5.2: Possible timing of water interventions

Options	Rapid-onset emergency		
	Immediately after	Early recovery	Recovery
1.1 Water points: changing management	✓	✓	✓
1.2 Water points: rehabilitating	✓	✓	✓
1.3 Water points: establishing	—	—	✓
2. Water trucking	✓	—	—

Options	Slow-onset emergency			
	Alert	Alarm	Emergency	Recovery
1.1 Water points: changing management	✓	✓	✓	✓
1.2 Water points: rehabilitating	✓	✓	✓	✓
1.3 Water points: establishing	✓	✓	—	✓
2. Water trucking	—	—	✓	—

Links to other LEGS chapters and other HSP standards

The provision of water is complementary to other livestock-based emergency responses. This is particularly the case for livestock feed interventions and livestock offtake. During livestock offtake, some animals are taken out of the production system. Water and feed are then provided to help ensure the survival of the remaining stock. Coordination between initiatives and among agencies is paramount to avoid one activity undermining another. Coordination with water requirements for crop production may also be needed (see SEADS).

The provision of water for livestock may also be complementary to human water provision. This is particularly so where the rehabilitation or establishment of water sources provides water of a suitable quality for both animals and humans. In contrast, however, there may be times when water trucking for livestock may compete with human water supplies, unless carefully managed. For further information on human water supplies, see the Sphere Handbook chapter on 'Water Supply, Sanitation and Hygiene Promotion.' Coordination is also necessary to avoid livestock contamination of human supplies. The need to ensure coordination between human and livestock water supply may become particularly important in camp settings, where space and water sources may both be limited *(see Chapter 7: Shelter)*.

LEGS Principles and other issues to consider

Table 5.3: Relevance of the LEGS Principles to water interventions

LEGS Principle	Examples of how the principles are relevant in provision of water interventions
1. Supporting livelihoods-based programming	Water supply is critical for protecting livestock assets from mortality. Productivity of livestock is maintained only through regular water intake, which ensures the productivity of livestock livelihoods.
	Dysfunctional or disrupted water supplies resulting from emergencies affect the income and livelihoods of at-risk families. This is true in urban or rural settings, and pastoral or smallholder production systems.
2. Ensuring community participation	Active community participation helps to avoid the risks associated with marginalisation (e.g. when richer groups secure private means of water provision for their animals). It thereby ensures equitable access to water for all social groups.

continued over

Technical standards
Provision of water

LEGS Principle	Examples of how the principles are relevant in provision of water interventions
2. Ensuring community participation *(continued)*	Water management committees can help identify appropriate solutions for adequate water supply. Committees should include all social, gender and wealth groups, and those who control private water businesses. They can ensure fair access for all subsets and at-risk groups of a crisis-affected community. Committees can also lead to cost sharing of interventions *(see Appendix 5.3)*.
3. Responding to climate change and protecting the environment	Excessive water extraction should be avoided as much as possible. Establishing new permanent water points must be carefully planned to avoid contributing to environmental degradation.
	Preparedness activities should include proper technical, economic and social assessment of water supply options, with a view to ensuring water development plans take account of climatic trends.
4. Supporting preparedness and early action	Water point rehabilitation as part of preparedness activities supports communities likely to be impacted by emergencies.
	Integration of emergency water strategies into long-term water development plans, co-developed with local government, can take disaster risk reduction (DRR) into account.
	Guidance/training on management of water sources may help a community to be better prepared for emergencies.
5. Ensuring coordinated responses	Water for livestock may be provided at the same time as other interventions, such as feed provision, veterinary support or livestock offtake. Coordination is important to ensure that the activities complement and do not undermine each other.
	Coordination is particularly important in the provision of water for livestock and humans to ensure the needs of both are met. To limit the spread of disease, including zoonotic threats as well as transmission from/to wild species, livestock must not contaminate water supplies for humans.

continued over

Technical standards
Provision of water

LEGS Principle	Examples of how the principles are relevant in provision of water interventions
6. Supporting gender-sensitive programming	Livestock are important for livelihoods of both women and men. Therefore their participation in all water-related management at community level can positively contribute to appropriate responses and ensure fairer water distribution.
	Long distances to water points present potential risks and threats for women and children, in particular, through exposure to violent assault. These should be avoided through protection mechanisms (e.g. accompanied water fetching, lighting, guards, etc.).
	Gender roles may have changed during an emergency or crisis, particularly for poorer community members (e.g. because of more limited access to water). These changes and implications need to be understood in order to address potential inequities.
7. Supporting local ownership	Strengthening local ownership of humanitarian responses includes identifying existing structures, such as water management committees and other local initiatives linked to water provision, emergency response and preparedness.
	Land rights, ethnicity, and local politics may all affect access to water. Using local knowledge of water point management and conflict avoidance is particularly important within communities affected by complex or recurring crises.
	Customary knowledge of the relationship between water sources and natural resource management, and other relevant cultural practices, are best understood through local leadership.
	Local water committees and community cash contributions can help with the establishment, repair and management costs of any water resource over the longer term.

continued over

Technical standards
Provision of water

LEGS Principle	Examples of how the principles are relevant in provision of water interventions
8. Committing to MEAL	Monitoring water schemes is important to ensure that equitable access is maintained for at-risk groups, and that costs remain affordable. In the same way, monitoring the water management system will enable implementing agencies to address issues and challenges.
	Evaluating how schemes ensure water shortage recovery can lead to the strengthening of longer-term resilience. Sharing learning of intended and unintended consequences with humanitarian and development actors in the water sector helps build accountability and long-term strategy development.

Security and conflict

Personal security and protection of water users are important considerations in all emergency water options *(see Chapter 1: Box 1.3, Protection Principles)*. For example, during complex emergencies, people watering animals at water points may be at risk from livestock theft, robbery, or attack, especially women. The water points themselves can be vulnerable and highly sensitive to attack, and must occasionally be specifically protected. In conflicts, wells and watering points can be deliberately contaminated with toxins or animal carcasses; and boreholes and deep wells intentionally filled with rocks.

Many natural water points have pre-existing customary institutions responsible for their management. If interventions fail to involve existing water management structures in emergency response, this may cause friction between existing and new water users. Potential issues must be identified prior to rehabilitation or establishment of water points to avoid ownership conflicts. This will also ensure equitable access and sustainable systems for the future.

Issues of water conflict are particularly sensitive in settlements and camps *(see Chapter 7: Shelter)*. Camp residents who need to access water points outside the settlement for their livestock may come into conflict with host populations. Early negotiation with all stakeholders can help to minimise potential conflicts.

Technical standards
Provision of water

Decision tree for water provision options

The decision tree *(Figure 5.1)* summarises some of the key questions to consider in determining the most feasible and appropriate option for an emergency water provision intervention. The standards, key actions and guidance notes that follow provide more information for detailed planning. Where possible, they build on preparedness activities conducted prior to the onset of the emergency/in 'normal' times.

Technical standards
Provision of water

Figure 5.1: Decision tree for water provision options

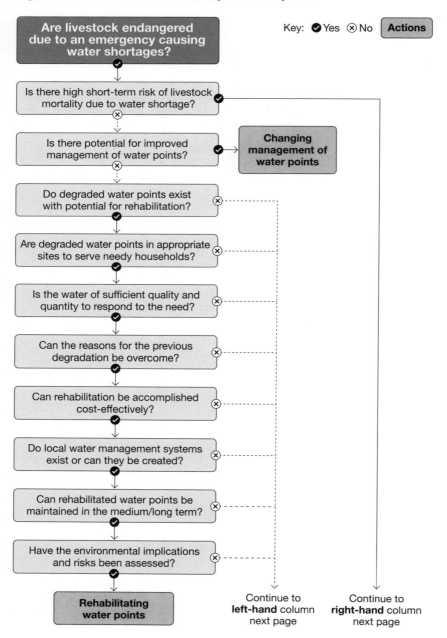

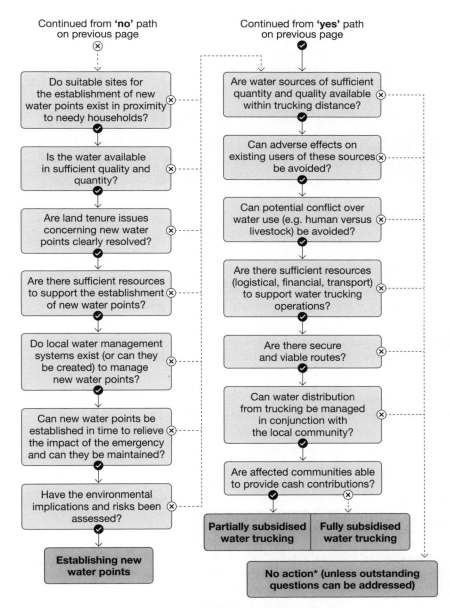

*The result **'No action'** does not necessarily mean that no intervention should take place, but rather that further training or capacity strengthening may be required in order to be able to answer 'yes' to the key questions.

Technical standards
Provision of water

The standards

Before engaging in water provision initiatives, the feasibility and appropriateness of the different technical options should be carefully considered, as highlighted in *Figure 5.1*.

> ### Standard 1: Preparedness
> Preparedness activities facilitate planning for sustainable and appropriate water provision for livestock in emergencies.

Key actions

- Map the location of existing and non-functioning water sources and note their capacity *(see Guidance note 1)*.
- Identify existing water management systems *(see Guidance note 2)*.
- Assess the current and potential environmental impact of continued or increased water extraction from existing water sources *(see Guidance note 3)*.
- Identify existing water development initiatives and the potential for integrating emergency water strategies into ongoing longer-term activities *(see Guidance note 4)*.

Guidance notes

1. Mapping of existing water sources and capacity

Information on existing (and non-functioning) water sources should be gathered, including location and capacity, in 'normal' times, prior to the onset of any emergency. Where possible, information should also be gathered on water quality. Agencies on the ground may already have this information, particularly those working on longer-term water development. The availability of this information facilitates rapid emergency assessment and response planning *(see Standard 2)*.

2. Analysis of existing water management systems

Water is a common natural resource, and its utilisation for humans, animals and crops requires clear management practices. These include ensuring that time of use, equal water distribution, maintenance and hygiene measures are all adhered to.

Boreholes, ponds, dams and reservoirs, as well as shallow and deep wells, are usually managed by local (often customary) institutional arrangements. Emergency water provision should recognise and build on existing management systems wherever possible and appropriate. Therefore, prior identification is vital to inform emergency response planning.

In areas prone to conflict and social tensions, preparedness should involve an analysis of whether water access is a factor in exacerbating conflict. Prior conflict analysis can also identify options for mitigation, such as community-based conflict resolution mechanisms.

3. Environmental impact

Where heavy use of a water source is already causing negative environmental impact, any additional use during an emergency will exacerbate this situation. Prior understanding of vulnerable areas can inform planning to ensure that negative environmental impacts of emergency water provision are minimised.

The assessment should consider the impact on the environment of the location and capacity of any potential new water source. The siting of new water sources can have a very negative environmental impact. Conversely, when water points are planned in conjunction with natural resource management strategies, the impact on the environment and on the natural resources available for livestock can be beneficial.

4. Water development

The development of water sources, in particular the construction or rehabilitation of infrastructure, is generally a longer-term development activity. Where possible, development agencies can build contingency plans for emergency water provision into these longer-term programmes. These can include budget provision or flexible funding options to allow emergency water interventions to be planned and implemented rapidly. Water point development can provide an opportunity for greater social cohesion between different groups (such as pastoralists and crop-based farmers), as they collaborate on a common output for their shared water needs.

**Technical standards
Provision of water**

> ## Standard 2: Assessment and planning
> Water provision for livestock is based on an analysis of needs and opportunities, as well as existing local water management systems.

Key actions

- Building on any preparedness activities *(see Standard 1)*, conduct an analysis of different options that can be used to form the basis for water provision *(see Guidance note 1)*.
- Assess existing and degraded water sources for water quality *(see Guidance notes 2 and 3)*.
- Identify existing or new effective management systems. This will ensure continued provision of water of acceptable quality without conflict, while addressing the needs of all the different users, including at-risk groups *(see Guidance note 4)*.

Guidance notes

1. Analysis of options using existing water sources

The planning of water provision activities should begin with a participatory assessment of existing water sources to review the availability, accessibility, affordability and quality of the water they provide. This participatory assessment should include analysis of water sources that have fallen into disrepair and are no longer used. The assessment should take account of water access for poorer and more at-risk households. Where preparedness planning is possible, this information may already be available *(see Standard 1)*.

The needs for a water supply for humans should also form part of this initial analysis.

When considering the option of relocating livestock to other existing water sources, analyses should consider the cost, amount of water (and feed) available, and the capacity and willingness of the current water users to absorb additional animals.

Because the cost of trucking water from other water points is very high, assessments should explore other existing options first before considering this option. *Appendix 5.1: Assessment checklist for water points* contains a checklist for assisting with rapid water point assessment.

2. Water quality for livestock

Water quality for livestock is generally a much less critical issue than water quality for humans. However, animals can also be affected by high mineral content and water-borne diseases such as salmonellosis, anthrax and colibacillosis. In the absence of a recognised field test to assess the bacterial content of water, a basic investigation is recommended. This should look at possible chemical contamination (nearby factories) and bacteriological/organic contamination (human settlements), including consultation with the local community (see *Appendix 5.1: Assessment checklist for water points*; see also 'water quality' in *FAO* (2016, p. 112)).

3. Contamination of human water sources

Where livestock and humans share water sources, the water may become contaminated by animals and affect human health and well-being. Simple management measures can be used to prevent this, including the use of troughs or pans for livestock watering. Protection of water sources may also be necessary to prevent the water becoming contaminated with acaricides and other chemicals used on livestock.

4. Water management systems

Many customary practices exist for regulating water access rights and water point care in 'normal' times *(see Standard 1, Guidance note 2)*. Any rehabilitation of existing water sources, or the establishment of new sources, should take into account these management systems. However, they may not be sufficient to cope with the challenges of rapid-onset, and also slow-onset, emergencies. Agencies may need to provide inputs and support from outside.

In some societies, social constraints may make it difficult for different ethnic or caste groups to access the same water point. These issues need to be handled with considerable sensitivity to ensure response options guarantee equitable access for all. LEGS *Principle 6: Gender-sensitive programming*, is specifically recommended to strengthen sustainable and equitable water use. Participation of women in the decision-making process is known to increase community inclusiveness and contribute to the protection of the rights of at-risk groups.

Technical standards
Provision of water

Standard 3: Location of water points

Water source rehabilitation and establishment interventions are carefully located to ensure equitable access to water for the livestock of the most at-risk households in the emergency-affected areas.

Key actions

- Base the selection of water points for rehabilitation and/or the location of sites for establishing new water points on a sound assessment of current and future demands. This should take account of the demand from both local human and livestock populations (see Guidance note 1).
- Ensure that the capacities of the water sources to be used can be reasonably expected to meet needs throughout the period of the emergency and beyond (see Guidance note 2).
- Take insecurity implications into account when planning the location of water points (see Guidance note 3).
- Organise siting of water points in conjunction with (male and female) community leaders, preferably building on existing customary water management systems (see Guidance note 4).

Guidance notes

1. Assessment of demand for water

An assessment of demand for water during the emergency should be made based on the best estimates derived from livestock population censuses, local authority records, and consultation with local affected communities. The assessment should consider ease of access for livestock and for water collection. For example, if livestock are to consume water at the water point, then the demand assessment should consider reasonable walking distances to determine the area to be covered by the water point. It should also look at numbers of troughs, drinking schedules, watering intervals, holding areas and expected waiting times. Where water will be carried to where the animals are located, similar assessments should be made.

2. Ensuring adequacy of the water supply

The supply from a water point may be inadequate for meeting projected demand. In this case, additional arrangements may be necessary (for example, establishing additional water points nearby or trucking in extra supplies). The assessment of the adequacy of all proposed water supplies should take into account the likely future utility of the water points. This should be both generally and in the event of other emergencies.

3. Insecurity

People taking their animals to water points may be at risk from livestock theft, general robbery, and other forms of personal attack. This is because their movements are easily predicted. The security needs of women and children in these situations are particularly important. Liaison with agencies responsible for security in affected areas is needed when determining water point locations, so they can anticipate these dangers as much as possible. They can then put conflict-prevention and other safety measures in place.

4. Community leadership

As highlighted in *Standard 1: Preparedness* and *Standard 2: Assessment and planning*, local water management systems should be consulted when siting water points. This should apply both to the rehabilitation of existing sources and the establishment of new ones. Community leadership is vital to ensure the future management and maintenance of the water source beyond the emergency. It will also encourage sustainable and equitable access to water for all community members. This may be particularly important in camps because of potential competition for the resource between camp residents and the local population. In these situations, negotiation and agreement with community leaders is paramount to avoid conflict.

> **Standard 4: Rehabilitation and establishment of water points**
>
> Rehabilitated or newly established water points represent a cost-effective and sustainable means of providing clean water in adequate quantities for the livestock that will use them.

Key actions

- Consider the rehabilitation of water points as an intervention only when demand in the affected area cannot be adequately met by extending the use of existing water points *(see Guidance note 1)*.
- Undertake a full survey of degraded water points and the reasons for the degradation for all locations in the affected area where demand exists, or is likely to develop *(see Guidance note 2)*.

Technical standards
Provision of water

- Consider establishing new water points as an intervention only when the use of existing water points or their rehabilitation is not possible and when the consequences have been carefully considered *(see Guidance note 3)*.
- Deliver to the selected locations the technical inputs and materials required to implement the rehabilitation/establishment programme effectively *(see Guidance note 4)*.
- Consider mechanisms that promote cost sharing between external agencies and affected communities *(see Guidance note 5)*.
- Ensure that people are available (and trained) for the routine management and maintenance of water points *(see Guidance note 6)*.
- Ensure that sustainable management structures are in place, supported by community leadership *(see Guidance note 7)*.

Guidance notes
1. Confirmation of the need to rehabilitate water points
Extending the use of existing water points for communities affected by the emergency is a cheaper option than water point rehabilitation. However, assessments should carefully evaluate the potential for introducing conflict between existing and new users at the planning stage. In practice, it may be possible to offer some coverage of affected communities by using existing sources, but this may need to be boosted by rehabilitation as well.

2. Identification of water points suitable for rehabilitation
A comprehensive survey is a first requirement for developing a cost-effective programme of water point provision. For each water point, this should include details on:

- water quality;
- the resources required to undertake a rehabilitation programme;
- likely water capacity (quantity and continuity);
- extent of damage and ease/cost of repairs;
- demand from users;
- why the point has become degraded, and the implications for its successful rehabilitation (issues such as conflict, water quality, and confusion over ownership may contribute to lack of use, as well as technical and maintenance issues).

3. Confirmation of the need to establish new water points

Where rehabilitation of existing water points will not offer adequate coverage for the affected community, the intervention may need to establish new ones. The potential consequences of establishing a new water point include land tenure issues, modification of grazing patterns, environmental degradation, and competition over resources. These should be carefully analysed.

4. Technical feasibility of the options

Understanding the reasons why water points have fallen into disuse can inform the assessment of the technical feasibility of rehabilitation. Other basic requirements include:

- availability of qualified water engineers and labourers to implement interventions;
- capacity to deliver required materials to the site, including adequate access roads;
- continuous availability of spare parts for wells and boreholes.

The equipment required for establishing new water points is likely to be considerably heavier than that for rehabilitation (for example, drilling rigs/excavation equipment for digging wells). It may therefore require higher-capacity transport and better roads to allow access.

5. Cost recovery

As part of the assessment process, there should be an analysis of community capacity and willingness to contribute to the emergency response water scheme. It may be necessary to convince decision makers, traditional elders, political leaders, and particularly influential female leaders to encourage community contributions in emergencies.

6. Responsibilities

Water points need routine management and maintenance, as well as people (whether community members or agency staff), for:

- routine checking to ensure that water quality and supplies are being maintained;
- monitoring to ensure that access is maintained equitably for all users;
- resolution of disputes among different user groups;
- routine maintenance and ordering and replacement of damaged parts (manual wells are generally less damage-prone than boreholes);

Technical standards
Provision of water

- appropriate training of water committees for local users (taking into consideration that men are often the gatekeepers of customary practices that can limit women's access to both resources and public roles).

7. Management structures

Whether water points are community-managed or privately owned, it is important to ensure that sustainable management structures are sustained/established, with the endorsement of community leadership. This is vital for the future management and maintenance of the water source beyond the emergency. It will also contribute to sustainable and equitable access to water for all community members *(see Appendix 5.3)*.

> **Standard 5: Sources and quality for water trucking**
>
> Water for trucking is obtained from sources that can maintain an adequate supply, of assured quality, during the planned intervention period.

Key actions

- Implement water trucking only as a short-term measure when other options are not possible *(see Guidance note 1)*.
- Ensure the supply of water can be maintained throughout the lifespan of the proposed trucking operations *(see Guidance note 2)*.
- Ensure that use of water sources by trucking operations does not compromise the needs of their existing users and has the approval of any relevant statutory authorities *(see Guidance notes 2 and 3)*.
- Ensure that the use of water sources does not reduce the availability of water for human populations *(see Guidance notes 3 and 4)*.
- Ensure that the water quality is suitable for livestock *(see Guidance note 5)*.
- Ensure that tankers and other water containers are properly cleaned before use *(see Guidance note 6)*.

Guidance notes

1. Short-term measure

Water trucking should be considered as a last-resort measure to save the lives of livestock as it is expensive and administratively complicated. Using trucks to deliver water for human use is also generally discouraged. Other options, including relocation of livestock closer to existing sources of water, should be thoroughly explored before selecting trucking as an option.

2. Continuity of supply

Although water trucking operations should aim to operate only in the short term, this is not always possible. Whatever the term of the operation, a realistic assessment of the continuity of water supplies needs to be made at the planning stage. This assessment should:

- Assess whether proposed water sources have the physical capacity to continue to provide supply during the operation. The potential for selected sources to be affected by the spread of the emergency should be considered.
- Secure permission from existing users or from the relevant authorities to access the source.
- Ascertain whether accessibility of the sources can be maintained; for example, repeated passage of trucks might degrade access routes.
- Consider budgetary implications carefully, as water trucking is generally a high-cost operation. Operational budgets need to be sufficient to handle extended trucking services if alternative interventions are delayed. Costs can be significantly reduced if water sources can be located close to the ultimate distribution points. However, this can increase the risk of conflict with existing users or threats to the continuity of supply.

3. Considering the needs of existing users

Water sources used for trucking operations are likely to have existing users. Conflict with existing users can seriously undermine the viability of the operation and create more adversely affected communities. Locating water sources as close as possible to where the water will be consumed may be financially desirable. However, this should not mean compromising these sources for their existing users. During the planning stages of a trucking operation, managers need to engage with local leaders and other stakeholders and, where possible, use local mediation procedures to ensure that existing users' needs are properly taken into account.

4. Coordinating with the demands of human populations

In situations where water is scarce or where resources for trucking operations are limited, the immediate needs of human populations must always be prioritised. However, meeting the demands of human and livestock populations does not have to be exclusive. In the case of a widespread emergency, the trucking infrastructure may be inadequate to service both people and animals. However, small-scale localised operations may be able to deliver an integrated service that supplies water to people and their livestock. Provided that the availability of trucks and staff is adequate, water for livestock may be derived from sources that are not of sufficient quality for consumption by humans.

5. Water quality

In cases where water trucking is for both humans and livestock, the Sphere standards for water quality apply. However, if high-quality water sources are limited, interventions may source poorer-quality water from rivers or standing lake water, which cannot feasibly or economically be purified for human consumption, for use by livestock only.

6. Cleanliness of tankers

Tankers may have been used for transporting other types of liquid, including toxic pesticides, herbicides, solvents, fuels, and sewage. Unless their previous history is reliably known, all vessels and distribution equipment should be thoroughly cleaned and disinfected before being released for use in water trucking operations.

Standard 6: Logistics and distribution for water trucking

Water is transported securely and distributed equitably in the affected area.

Key actions

- Ensure adequate staff capacity is retained during the intervention, through effective investment and management *(see Guidance note 1)*.
- Ensure that adequate resources are available to meet the recurrent costs of fuelling and servicing the tanker fleet and associated equipment *(see Guidance note 2)*.
- Where possible, select routes that will not be degraded by the frequent passage of heavily laden water trucks *(see Guidance note 3)*.
- Set up distribution points in appropriate locations, and accommodate any livestock movements that may occur during the operation *(see Guidance notes 4 and 5)*.
- Undertake proper security assessments for the proposed water distribution *(see Guidance note 6)*.

- Review the ability of the community to contribute and the need for subsidies *(see Guidance note 7)*.

Guidance notes
1. Staffing
Successful trucking operations require consistent and sustained staff inputs, notably competent and experienced managers and supervisors. It is also important to ensure that drivers and assistants are kept motivated through proper reimbursement and careful attention to other needs, including subsistence allowances and personal security.

2. Maintenance and fuel supplies
Qualified mechanics and reliable supplies of fuel and equipment need to be available throughout the duration of the trucking operation. This includes any material needed to operate and maintain pumps, containers and delivery equipment. Major issues to consider include:

- **The cost and availability of fuel:** Ideally, it should be possible for drivers to refuel without making major detours from the trucking route. This may require fuel to be brought in separately, adding to the logistical complications of the operation. This should also be a consideration in the original selection of water sources.
- **Spare parts:** These should be readily obtainable; locally made equipment that is easily repairable is preferred.

These issues (particularly those relating to maintenance) may affect the decision regarding the type of transport that will be used by the trucking operation (for example, trucks or tractor trailers with bowsers or bladder tanks).

3. Ensuring the integrity of supply routes
Supply routes should be adequate for the passage of laden water tankers. Otherwise, the response plan will need to make provision for their maintenance and repair: for example, through cash for work schemes. In addition, communities living along the access route must also have their water needs addressed and be made aware of and approve the plan. If not, they may cause disruption by blocking the road or forcibly diverting the water to fulfil their own needs. Implementing agencies must evaluate and resolve any potential risk of conflict over the response in advance, to ensure that 'do no harm' principles are followed (see *Chapter 1: Introduction to LEGS* on the Sphere Protection Principles).

Technical standards
Provision of water

4. Managing distribution points

Livestock keepers may collect water from distribution points to take to their livestock or bring their animals to receive water directly, from a tank or pond. In either case, a system needs to be established to ensure the needs of all attendees are met equitably and sustainably, based on appropriate existing local water management systems. Where it is possible to establish storage facilities, trucking can be more efficient. This is because tankers can decant the water quickly and return to the source to collect more, thus reducing the waiting time.

5. Water trucking to mobile livestock

Relocation of livestock is often a customary response to drought. Where this occurs, trucking of water may be considered to support the migration. This will add considerably to the already complex logistics of water trucking.

6. Establishing a safe distribution network

The risk to the personal safety of staff employed in transporting water for use in emergency programmes should always be of paramount importance.

7. Full or partial subsidy

As part of the assessment process, there should be an analysis of community willingness and ability to contribute to the costs of water trucking. It may be necessary for agencies to either fully or partially subsidise water trucking.

Appendices

Appendix 5.1: Assessment checklist for water points

This checklist summarises the issues that need to be considered when assessing potential water points for use by livestock keepers in an emergency. Sources of information for answering the questions in this checklist may vary from rapid field assessments to, in principle at least, laboratory analyses for water quality. They should, however, always include some consultation with the different stakeholder groups in the local area. When assessments are carried out under time pressure, users may have to prioritise which questions need to be covered.

Supply of water

- Is the water point currently producing water?

If yes:

- Is the water point at risk of drying up over the course of the emergency response?
- What is the capacity of the water point to support the local livestock population?

If no:

- Is it technically feasible (both in terms of cost and timescale) to rehabilitate the water point to meet the needs of the local livestock population?
- Are there personnel available to manage and implement rehabilitation of the water point?

Accessibility

- Is the water point within easy reach of a significant population of affected livestock?
- Are there any social, cultural or political constraints to the use of the water point by livestock?
- Can water from the source be made available to affected livestock keepers in an equitable manner (regardless of age, gender, ethnicity or wealth)?
- Can affected livestock make use of the water point without:
  a. compromising the needs of existing users (human or animal)?
  b. risk to the personal safety of the livestock keepers (including women and children)?

Technical standards
Provision of water

  c. interfering with other aspects of the emergency response?

Water quality

- Are testing facilities (either field or laboratory) available to assess the adequacy of water quality for the source?

If yes:

- Is there access to laboratories that can analyse for major chemical contaminants?
- Are water testing kits available that can be applied to the water points/sources under consideration?
- Are suitably qualified technicians available locally to undertake assessments of microbiological contamination of water sources?

If no, the following questions may help make a rapid on-the-spot assessment:

- Is water from the source clear or cloudy?
- Is there any evidence of salinity in the area (for example, formation of salt pans)?
- Are there any local indicators of chemical contamination (for example, nitrates/nitrites, patterns of fertiliser and pesticide use, existence of local small-scale industries such as tanneries or light industries)?
- Have there been any reports of water-borne diseases from the source?

Note: Participatory methods and techniques that are particularly useful for water assessment include natural resource and service maps, and matrix scoring of different water facilities. Details on participatory methods are provided in *Appendix 3.1* of *Chapter 3: Emergency response planning* and the LEGS Participatory Techniques Toolkit.

Technical standards
Provision of water

Appendix 5.2: Examples of monitoring and evaluation indicators for water provision

Process indicators (measure things happening)	Impact indicators (measure the result of things happening)
Designing the intervention	
Number of meetings with community/community representatives and other stakeholders (including private sector suppliers where relevant)	Meeting reports with analysis of options for water provision
	Action plan including: roles and responsibilities of different actors; approach for water supply (e.g. rehabilitation of existing sources or establishing new sources); community involvement in managing rehabilitated or new water points
Provision of water	
Number of water points rehabilitated or constructed by type and location	Accessibility of water (physical distance and safe access to water) for users and their livestock, including vulnerable groups
Delivery capacity of water points	
Volume of water provided by trucking	Availability of water – sufficient for livestock needs
	Quality of water – suitability for livestock
	Number of livestock-keeping households using water points versus number of livestock-keeping households needing water; breakdown of figures by vulnerable group
	Number of livestock using water points by livestock type; frequency of watering
	Increase or decrease in women's and girls' labour burden to collect water for livestock

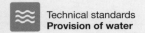
**Technical standards
Provision of water**

Appendix 5.3: Considerations for water point management

The establishment of clear management systems for water sources is important to secure the equal, fair and unbiased distribution of water. This will also help to keep a maintenance schedule in place. Experience has shown that unless management issues are considered at the beginning of an intervention, access may not only be inequitable but may result in conflict.

Management is also essential for the longer term. This is so that water points do not fall into disrepair after the emergency is over, but continue to operate into the recovery phase and beyond. Some potential issues for community management are covered below. Where water points are privately owned, the management arrangements will differ.

Diversity of membership: A management committee for a water point requires a diversity of members from various social and cultural levels within a community. If necessary, agencies should provide external support to local communities to strengthen water committee management systems.

Inclusion of female representatives: The involvement of women is specifically recommended to strengthen sustainable and equitable water management. It is also recommended that women are included on water management committees because they generally need to negotiate for domestic water use as well.

Water points and water trucking: The management of water distribution from water trucking activities can build on local water point management systems. This will help ensure equitable distribution and access within communities.

Transparency: When cost sharing is involved in emergency water responses, transparent procedures must be applied. By its nature, management of funds and cost contributions, whether in cash or in kind, is very sensitive. Financially responsible actors need to work closely with selected community members (and the organisation behind the intervention) to be able to disclose accounts and values at any time.

Community savings: If funds and scope allow, a community may consider a savings scheme and/or community bank account to prepare for a future emergency. It can define options for spending the funds depending on the type of emergency and immediate needs, with the guidance of specialists as appropriate.

References and further reading

Cash and cost recovery
Free online training modules on Water Trucking and Cost Recovery for Water Systems have been developed by DisasterReady/Mercy Corps, in English and Arabic, https://get.disasterready.org/water-trucking-and-cost-recovery-online-training/

ISW/SDC (2016) *How to establish a Full Cost Recovery Water Supply System? What are the Key Factors for Success and Replication?*, SDC Briefing Note, Federal Department of Foreign Affairs (FDFA), Bern, https://www.rural-water-supply.net/en/resources/details/770

Technical guidance for water emergencies
House, S. and Reed, R. (2004) *Emergency Water Sources – Guidelines for Selection and Treatment*, 3rd edn, Water, Engineering and Development Centre (WEDC), Loughborough, https://wedc-knowledge.lboro.ac.uk/details.html?id=18064

Water and livestock emergencies
FAO (2016) *Livestock-related interventions during emergencies – The how-to-do-it manual*. Edited by Philippe Ankers, Suzan Bishop, Simon Mack and Klaas Dietze. FAO Animal Production and Health Manual No. 18. Rome, https://www.fao.org/3/i5904e/i5904e.pdf (See Chapter 7)

Gebru, G., Yousif, H., Adam, A.E.M., Negesse, B. and Young, H. (2013) *Livestock, Livelihoods, and Disaster Response, Part Two: Three Case Studies of Livestock Emergency Programmes in Sudan, and Lessons Learned*, Feinstein International Center, Tufts University, Medford, MA, http://fic.tufts.edu/publication-item/livestock-livelihoods-and-disaster-response-part-two-2/

Water trucking
Coerver, A., Ewers, L., Fewster, E., Galbraith, D., Gensch, R., Matta, J., Peter, M. (2021) *Compendium of Water Supply Technologies in Emergencies*, 1st edn, German WASH Network (GWN), University of Applied Sciences and Arts Northwestern Switzerland (FHNW), Global WASH Cluster (GWC) and Sustainable Sanitation Alliance (SuSanA), Berlin, ISBN: 978-3-033-08369-1, https://www.washnet.de/wp-content/uploads/2021/09/GWN_Emergency-Water-Compendium_2021_new.pdf

WEDC, Loughborough University (2011) *Delivering safe water by tanker*, WHO Technical Notes on Drinking-Water, Sanitation and Hygiene in Emergencies, Number 12, WHO, Geneva, https://wedc-knowledge.lboro.ac.uk/resources/e/mn/042-Delivering-safe-water-by-tanker.pdf

Wildman, T., Brady, C., and Henderson, E. (2014) 'Rethinking emergency water provision: Can we stop direct water trucking in the same places every year?', *Humanitarian Exchange Magazine*, issue 61, article 13, Humanitarian Practice Network, https://odihpn.org/publication/rethinking-emergency-water-provision-can-we-stop-direct-water-trucking-in-the-same-places-every-year/

Wildman, T. (2013). *Technical Guidelines on Water Trucking in Drought Emergencies*, Oxfam GB, London, https://policy-practice.oxfam.org/resources/technical-guidelines-on-water-trucking-in-drought-emergencies-301794/

See also case studies for water interventions in emergencies at: https://www.livestock-emergency.net/resources/case-studies/

Chapter 6: Technical standards for veterinary support

- 207 **Introduction**
- 208 **Options for veterinary support**
- 216 **Timing of interventions**
- 217 **Links to other LEGS chapters and other HSP standards**
- 218 **LEGS Principles and other issues to consider**
- 223 **Decision tree for veterinary support options**
- 226 **The standards**

- 241 **Appendix 6.1: Assessment methods and checklist for veterinary support**
- 243 **Appendix 6.2: Examples of monitoring and evaluation indicators for veterinary support**
- 245 **References and further reading**

Photo © FAO/B. Geers

Chapter 6: Technical standards for veterinary support

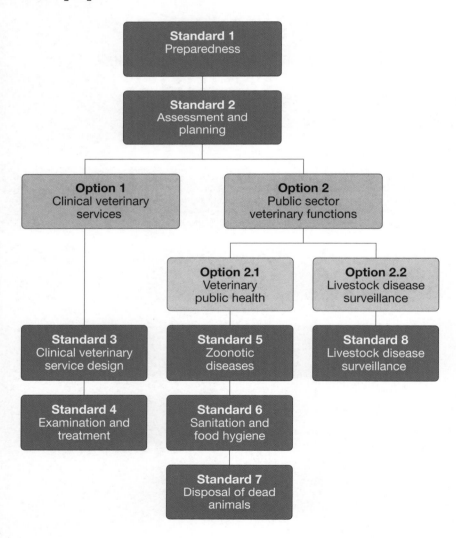

Technical standards
Veterinary support

Introduction

Livestock keepers often describe veterinary support as critical for protecting livestock in emergencies, together with providing feed and water. Veterinary care such as early diagnosis and treatment can help to prevent the death of animals from disease. Emergencies that lead to high livestock mortality have a major impact on livelihoods as it may take years for livestock-keeping communities to recover. In many situations, livestock keepers aim to protect core breeding animals. In general, veterinary vaccines and medicines are inexpensive items relative to the value of livestock that would otherwise be lost.

Agencies can support private veterinary service providers, NGOs and governments to provide clinical veterinary services. A veterinary intervention that works with local service providers during emergencies will help maintain the supply of livestock products for affected communities. Interventions can also help ensure that veterinary services are available in the post-emergency phase. Clinical veterinary care is usually the priority response during emergencies. However, in some contexts, this might be complemented with support to public sector veterinary functions such as veterinary public health and livestock disease surveillance.

This chapter presents information on the importance of veterinary support in emergency response, together with the technical options for veterinary support and the associated benefits and challenges of each. Information is also available in Chapter 5, 'Veterinary support', in *FAO* (2016). For each technical option, LEGS provides information through standards, key actions and guidance notes. Checklists for assessment, as well as monitoring and evaluation indicators, are presented as appendices at the end of this chapter. Further reading is also provided. Case studies are presented on the LEGS website (see https://www.livestock-emergency.net/case-studies/).

Links to the LEGS livelihoods objectives

Providing veterinary support in an emergency helps achieve two of the LEGS livelihoods objectives, namely **to support crisis-affected communities to**:

- **protect key livestock assets**; and/or
- **rebuild key livestock assets**.

Besides preventing the death of livestock, veterinary care has positive impacts on livestock production and welfare. This care can increase benefits derived from animals, whether from milk production, fertility or their use as working animals. Veterinary care can also help to rebuild valuable livestock

Technical standards
Veterinary support

assets, whether these consist of pastoralists' herds, a single donkey, a pair of draught oxen or just a few chickens.

The importance of veterinary support in emergency response

Livestock diseases are among the main causes of livestock mortality and reduced production in low- and middle-income countries. Emergencies can increase disease risk, with different kinds of emergencies having varying impacts on animal health. For example:

- Droughts, floods and harsh winters reduce access to grazing, resulting in weaker animals with lower capacity to withstand disease.
- Flooding displaces topsoil, creating favourable conditions for diseases such as anthrax, while floodwater creates conditions for vector-borne diseases such as Rift Valley fever.
- Natural hazards such as earthquakes and volcanic eruptions can cause direct injury to animals.
- The risk of infectious disease transmission increases when emergencies cause livestock from different areas to congregate together.
- Displacement of people and their livestock can lead to disease outbreaks if animals are moved to places where they have not acquired immunity to local diseases.
- The risk of zoonotic disease transmission between animals and people can increase in crowded camp conditions.

LEGS does not cover the prevention and control of epidemic livestock diseases of major international importance. Specific guidelines are available from the World Organisation for Animal Health (WOAH) and the Food and Agriculture Organization of the United Nations (FAO), as indicated in *Chapter 1: Introduction to LEGS*.

As part of emergency response, veterinary support contributes directly to two of the five animal welfare domains described in *Chapter 1*, namely 'health' and 'mental state'.

Options for veterinary support

LEGS recommends two main types of veterinary support for emergency response: **clinical veterinary services** and **public sector veterinary functions**. These options for veterinary support are not exclusive, and more than one option or sub-option may be selected and implemented. In line with

LEGS *Principle 2: Community participation*, veterinary interventions should be identified and prioritised with community members to ensure poorer and more at-risk livestock keepers are included. This will also ensure that interventions address the needs actually experienced by the livestock keepers. The Core Humanitarian Standard on quality and accountability (CHS) requires resources to be used effectively, efficiently and ethically. For veterinary support, this means an appropriate distribution of resources across different activities, but with an emphasis on support that directly protects livestock assets. Further guidance is provided under *Standard 2: Assessment and planning*.

Option 1: Clinical veterinary services

Livestock keepers usually prioritise clinical veterinary services (comprising diagnoses, treatments and vaccinations) during an emergency. These services can be delivered through the government, NGOs or private veterinary service providers. In many lower-income countries, clinical veterinary services are in transition from public to private sector. Therefore, the growing private veterinary sector may be the main source of clinical veterinary care. However, veterinarians are often based in major urban centres or near more commercialised livestock farms. In remote areas, veterinary paraprofessionals may be the main service providers, along with traditional healers and various informal suppliers of veterinary medicines.

Veterinary paraprofessionals such as community-based animal health workers (CAHWs) are located in communities. They play an important role in supporting both clinical and public sector veterinary services and functions during emergencies. In complex emergencies and droughts, CAHWs have reduced livestock mortality and improved service delivery at a relatively low cost. When good practices are followed, CAHW systems respond well to the animal health priorities of livestock keepers, and receive strong local acceptance and support. Good practices include the use of participatory approaches in the selection, monitoring, appropriate supervision and, where possible, training of both female and male CAHWs. Connections between CAHWs and other animal health workers and drug/equipment suppliers are also important. In some countries, however, CAHWs have no legal basis to work, and other animal health service delivery mechanisms may need to be considered.

Technical standards
Veterinary support

In humanitarian crises, preventive and curative clinical veterinary interventions fall into two broad categories that can be implemented simultaneously:

- examination and treatment of individual animals or herds;
- mass vaccination or mass medication programmes.

1.1 Examination and treatment of individual animals or herds: This option allows for animals to receive treatments specific to the diseases present at the time of the emergency. It assumes that animals in different households or herds may have different diseases, and therefore allows for flexibility in the clinical care provided. In some countries, this approach is increasingly supported during emergencies by veterinary voucher systems. These systems are developed jointly by the community, private sector and government partners (see *Process case studies: Veterinary voucher scheme in Kenya* and *Veterinary voucher schemes in Ethiopia*). Similarly, non-veterinary emergency responses that provide cash, directly or indirectly, to households can enable people to pay for veterinary care from private workers. In addition to providing case-by-case clinical care, these approaches aim to avoid situations in which the free provision of medicines undermines existing private veterinary services.

More individual clinical care may also lessen the risk of drug resistance compared with the mass treatment of animal populations.

1.2 Mass medication or mass vaccination programmes: These programmes are widely used with the aim of preventing diseases in livestock populations during emergencies. Most commonly, emergency mass medication or vaccination programmes are one-off events and are implemented at no cost to livestock keepers. Therefore, care is needed on the part of agencies to ensure that the financial viability of existing veterinary services is not undermined.

- Mass medication programmes often use anti-parasite medicines, especially for worms and ectoparasites such as ticks or lice. Practitioners and recipients of these widely used programmes have reported positive impacts. However, because research has indicated limited impact or cost-effectiveness (see *Impact case study: Limitations of mass deworming of livestock during drought in Kenya*), LEGS does not yet include a standard on mass medication. Should agencies choose the mass medication option, LEGS recommends proper evaluation to better document the impacts of mass medication and understand when and how it should be used. It is recognised that a particular challenge with evaluating mass deworming programmes is that some impacts may only be observed after the emergency. So this is something that needs factoring into the timing and design of evaluations.

- Mass vaccination programmes usually cover infectious diseases such as anthrax, clostridial diseases, forms of pasteurellosis, Newcastle disease, Rift Valley fever, and sheep and goat pox. Although widely used, evidence of the livelihood impact of mass vaccination during rapid-onset and slow-onset emergencies is very limited (see *Impact case study: Limitations of livestock vaccination during emergencies, Ethiopia*). Therefore, LEGS does not include a standard on mass vaccination. If agencies choose to support mass vaccination, LEGS recommends proper evaluation such as comparisons of mortality in vaccinated versus non-vaccinated livestock. Three specific technical issues concerning mass vaccination are:

    a. Timing: some types of vaccine will only be effective if they are administered before the period or season of high disease exposure. This timing may not coincide with the timing of the humanitarian crisis or response.

    b. Some livestock vaccines are inactivated (killed) vaccines that require an initial course of two doses to achieve protective immunity. In these cases, the administration of a single dose of vaccine can have minimal effect. Similarly, a 'booster' dose will only be effective following a complete initial course of vaccination, and the vaccination history of animals can be unknown.

    c. The proportion of animals to be vaccinated to achieve herd immunity varies by disease. These proportions are often not specified in national strategies for animal disease control.

In some situations, such as complex emergencies, LEGS supports vaccination as part of international disease eradication campaigns. In these cases, the vaccination strategy will be set by a lead technical agency such as the FAO, and should be followed.

Option 2: Public sector veterinary functions

In humanitarian crises, agency support to public sector veterinary functions may supplement weakened government capacity. Or it may intervene where no officially recognised government authority is present. It includes two broad types of activity:

- **veterinary public health**; and
- **livestock disease surveillance**.

Technical standards
Veterinary support

In complex emergencies, there are also examples of veterinary interventions supporting policy and institutional strengthening. This may relate especially to the use of veterinary paraprofessionals in emergencies.

When considering support to public sector veterinary functions, three important challenges are relevant in emergency contexts:

Prioritising veterinary support: Under LEGS *Principle 1: Livelihoods-based programming*, and LEGS *Principle 2: Community participation*, veterinary interventions should be identified and prioritised with communities. Such interventions should achieve one or more of the LEGS livelihoods objectives. There is a reasonable body of evidence showing how clinical veterinary care contributes to supporting livelihoods during emergencies. Yet there is far less evidence on the impact of support to public sector-related activities. Typically, livestock keepers prioritise clinical care of their animals.

Cost-effectiveness: Based on the CHS, resources should be used effectively, efficiently and ethically. Some interventions aim to support both clinical veterinary care and public sector veterinary functions. In this situation, limited guidance is available on the appropriate distribution of an intervention's human and financial resources across these two main types of activity.

Preparedness: For veterinary interventions, LEGS *Principle 4: Preparedness*, should focus on planning for the timely delivery of clinical veterinary care. This process can include developing inventories of local veterinary service providers; pre-positioning suppliers of good-quality medicines; and the provisional design of cash and voucher assistance (CVA) with communities, alongside local private sector and government partners. Planning can also involve ensuring that the veterinary workforce is prepared to participate in responses during emergencies. Preparedness does not include activities that are routine; core public sector activities during 'normal' periods such as policy reform on national animal health issues; or vaccination programmes.

2.1 Veterinary public health: Veterinary public health includes the understanding, prevention and control of zoonotic diseases, and human food safety. Zoonotic diseases are transmissible from animals to humans, either through animal-derived food such as meat or milk, or by contact with animals. In 'normal', non-emergency periods, the control of these diseases is a key public sector function. Zoonotic diseases include (for example) anthrax, salmonellosis, tuberculosis, brucellosis, rabies, mange, Rift Valley fever, and highly pathogenic avian influenza ('bird flu'). Specific guidelines for the prevention and control of these diseases are available from FAO and WOAH, including animal welfare considerations, as mentioned in *Chapter 1: Introduction to LEGS*.

Veterinary public health is multi-sectoral. It involves not only veterinarians and veterinary paraprofessionals in the public and private sectors, but also human health workers, agriculturalists, social scientists communication professionals and others. Closely related to veterinary public health is the concept of One Health. LEGS positions One Health concepts and activities under veterinary public health.

Veterinary public health also includes the safety of animal-derived foods like meat, milk or eggs. A specific concern is that some veterinary medicines leave residues in these foods, leading to possible consumption of residues by people. In humanitarian crises, the trade-offs between human food security and human food safety are not well understood. However, emergencies can occur in areas characterised by pre-existing high levels of human food insecurity and malnutrition. In some cases, malnutrition levels exceed the World Health Organization (WHO) cut-off for emergencies even in 'normal' periods. For people in this situation, the risk of continuing or worsening food insecurity far outweighs the risk of ill health due to consuming meat or milk that is contaminated with drug residues.

2.2 Livestock disease surveillance: Livestock disease surveillance aims to collect information on livestock diseases for the purpose of understanding the disease status of an area. It also supports the detection of new or emerging diseases. Surveillance systems can provide information on the prevalence and economics of diseases, and support programmes for the prevention or control of major infectious and zoonotic diseases. Information from disease surveillance can also support international trade in livestock and livestock products.

Examples of disease surveillance and investigation activities during humanitarian crises include:

- raising public awareness to stimulate disease reporting;
- training veterinary paraprofessionals to report disease outbreaks;
- supporting government surveillance systems by linking veterinary paraprofessionals' disease-reporting systems to official structures;
- facilitating timely and region-specific disease outbreak investigation and response;

- providing regular feedback in the form of disease surveillance summaries to the workers who report.

The benefits and challenges of the veterinary support options and sub-options are summarised in *Table 6.1*.

Table 6.1: Benefits and challenges of veterinary support intervention options

Sub-option	Benefits	Challenges
Option 1: Clinical veterinary services		
1.1 Examination and treatment of individual animals/herds	This allows flexibility and veterinary care on a case-by-case basis	If provided free, the coverage and duration of the service are likely to be limited by the budget
	It can support existing private sector service providers (e.g. through voucher schemes)	If provided free, it risks undermining existing private sector service providers
	It can involve the public sector (e.g. for supervision and quality control)	The quality of locally available medicines may be poor – pre-positioning or prior selection of suppliers of good-quality medicines is required
	Wide coverage is possible, particularly when well-trained and supervised veterinary paraprofessionals such as CAHWs are used	
	It allows targeted or strategic prophylactic treatment or vaccination of individual animals or herds at risk	
	Quantitative evidence of impact on animal mortality is available	
	There is a lower risk of environmental contamination relative to mass medication	
	There is a lower risk of contributing to drug resistance relative to mass medication	

continued over

Technical standards
Veterinary support

Sub-option	Benefits	Challenges
1.2 Mass medication or vaccination programmes	It is relatively easy to implement Mass deworming does not require a cold chain Cost per animal can be low Mass medication has the potential to provide income for the veterinary sector (for example, through voucher schemes)	There are weak laboratory facilities in many areas – especially during emergencies – for confirming disease diagnosis before targeting specific diseases Large-scale vaccination programmes are difficult to design properly without basic epidemiological information A single dose of inactivated vaccines as part of a one-off campaign may have limited effectiveness, depending on the disease in question Optimal timing of vaccination is before the high-risk period of exposure; this may not coincide with the humanitarian emergency Coverage is often determined by budget rather than technical design criteria Free treatment and vaccination can undermine the private sector For many vaccines, there is a need to establish or support cold chains There is a risk of poor immune response to vaccination in animals already weakened (e.g. due to lack of feed) The quality of locally available medicines may be poor – pre-positioning of suppliers of good-quality medicines is required

continued over

Technical standards
Veterinary support

Sub-option	Benefits	Challenges
Option 2: Public sector veterinary functions		
2.1 Veterinary public health	Public awareness raising is often inexpensive, although it may require specialised communication expertise to design and test educational materials in local languages	Effective approaches often need community participation; livestock keepers may have other priorities during emergencies
	This can foster collaboration between veterinary, human health, WASH and other sectors	If not carefully managed and timed, it can divert resources away from more direct livelihoods-based assistance
2.2 Livestock disease surveillance	This can complement all other veterinary interventions and assist impact assessment of these interventions	It needs to be based on clearly defined surveillance objectives
	It fosters linkages between the central veterinary authority and the affected area	It can easily become a data-driven rather than an action-oriented process
	It can help to promote international livestock trade in some countries and regions	If not carefully managed and timed, it can divert resources away from more direct livelihoods-based assistance

Timing of interventions

Support to clinical veterinary services is usually appropriate throughout each stage of rapid-onset and slow-onset emergencies. However, support to public sector veterinary functions may be most appropriate during the recovery phase. By this time, clinical veterinary care has addressed immediate threats from livestock diseases *(see Table 6.2)*.

Clinical veterinary care is usually appropriate throughout complex emergencies. The timing of support to public sector veterinary functions will vary by context. For example, there may be areas of an affected country with relatively low levels of conflict over long periods. This will allow a longer-term perspective to be developed, which might include support to veterinary public health or disease surveillance.

Technical standards
Veterinary support

Table 6.2: Possible timing of veterinary support interventions

Options	Rapid-onset emergency		
	Immediately after	Early recovery	Recovery
1. Clinical veterinary services	✓	✓	✓
2. Public sector veterinary functions	—	—	✓

Options	Slow-onset emergency			
	Alert	Alarm	Emergency	Recovery
1. Clinical veterinary services	✓	✓	✓	✓
2. Public sector veterinary functions	—	—	—	✓

If vaccination is being considered, the timing of vaccination can be critical. This is because many important livestock diseases have a seasonal occurrence. The duration of vaccine protection can also be limited, depending on the disease or vaccine in question. In these cases, vaccination should be conducted before the period of high disease exposure. However, this time of year may not coincide with the timing of the humanitarian crisis, and if so, vaccination may not be effective. In complex emergencies that last many years, vaccination calendars can be used to assist planning the correct timing of vaccination.

Links to other LEGS chapters and other HSP standards

Veterinary care alone does not guarantee livestock survival or productivity in emergency situations. So veterinary support should be integrated with other livelihoods-based livestock interventions. Livestock requires feed and water (see *Chapter 4: Livestock feed*, and *Chapter 5: Water)* and, in some areas, will require shelter (see *Chapter 7: Shelter)*.

Clinical veterinary services complement livestock offtake (see *Chapter 8: Livestock offtake)* by helping to ensure the survival of the remaining stock. Veterinary public health inputs, such as pre-slaughter and post-mortem examinations, are important for slaughter offtake. Additional veterinary

Technical standards
Veterinary support

support is required during the provision of livestock *(see Chapter 9: Provision of livestock)* to examine livestock before purchase, for example, and to deliver clinical services after livestock distribution.

Veterinary interventions should support and not undermine local service providers, including veterinary and paraprofessional workers, and private veterinary pharmaceutical suppliers. The Minimum Economic Recovery Standards (MERS) Handbook provides guidance for market analysis and implementing economic and livelihood programmes in humanitarian contexts.

LEGS Principles and other issues to consider

Table 6.3: The relevance of the LEGS Principles to veterinary support

LEGS Principle	Examples of how the principles are relevant in veterinary support interventions
1. Supporting livelihoods-based programming	Veterinary support can reduce livestock mortality and therefore protect livestock as financial and social livelihood assets.
	By working with local veterinary service providers and not distributing free medicines, agencies help to ensure they are not undermining the services needed after an emergency ends.
2. Ensuring community participation	Effective veterinary responses depend on the active participation of communities in response design and implementation. Participatory impact evaluation is recommended to understand the impact of livestock responses.
	Livestock keepers can make important intellectual contributions to service design, assessment and delivery. They often possess detailed indigenous knowledge about animal health problems, including disease signs, modes of disease transmission and ways of preventing or controlling diseases. This knowledge is particularly well documented for pastoralist and agro-pastoralist communities. Agency-provided training, and support for local people to become CAHWs, can and should build on this knowledge.

continued over

Technical standards
Veterinary support

LEGS Principle	Examples of how the principles are relevant in veterinary support interventions
3. Responding to climate change and protecting the environment	Using medicines on individual animals or herds selectively will lessen the risks of environmental contamination (for example, avoiding the large-scale application of certain acaricides or anthelmintics).
	The potential impact on the environment always needs to be considered in an emergency response, particularly in an emergency, such as drought, that severely impacts natural resources. Veterinary support is unlikely to result in herd sizes that cannot be locally sustained, and helps to maintain a sustainable population of healthier, more productive animals.
4. Supporting preparedness and early action	Effective veterinary support requires quality veterinary medicines. So preparedness is critical to identify appropriate quality sources, and to map and pre-position supply chains and storage facilities.
	Preparatory use of participatory epidemiology can identify and prioritise most diseases specific to common emergencies.
	Preparedness and/or early action on potential bureaucracy and ensuring effective funding flows are essential for rapid response.
5. Ensuring coordinated responses	Coordination can help to ensure that veterinary support is provided in a consistent manner. Coordination and agreement on roles and responsibilities for the delivery of veterinary support is important to avoid confusion.
	Other LEGS technical interventions need to be coordinated with veterinary support to maximise effectiveness.
	Harmonised approaches are needed across interventions that use CAHWs (for example, for CAHW selection, training and supervision).
	Harmonised approaches are also needed across veterinary interventions with a pricing or voucher value component to ensure consistency.
	Coordination can involve harmonising the timing of specific actions such as vaccination (for example, by using vaccination calendars in complex emergencies).

continued over

Technical standards
Veterinary support

LEGS Principle	Examples of how the principles are relevant in veterinary support interventions
6. Supporting gender-sensitive programming	Gender-sensitive programming within veterinary support is an opportunity to address women's frequent lack of access to veterinary staff or to the traders who sell animal health products. Animal health professionals should have the skills and knowledge for treating the working animals that are often used by women.
	Livestock extension workers and training programmes should have a gender component. Where possible and appropriate, they should involve women through specifically targeted training activities and by recruitment of women CAHWs.
	Gender-sensitive programming is important for identifying gender-specific issues relating to livestock keepers and livestock health in an emergency. Animal health officials need to prioritise the animals owned by women, such as poultry, small ruminants and donkeys.
	Female-headed households and at-risk groups may require specific veterinary support.
	Women often have significant ethno-veterinary knowledge that should be taken into account in response planning. Gender-sensitive programming can also identify where divisions of labour may have changed (for example, when women become responsible for different livestock species following an emergency).
7. Supporting local ownership	Interventions that provide support to clinical veterinary services can be more effective and sustainable if community-based approaches are supported. Such approaches will recognise local people's significant capacities for primary animal healthcare.
	Community-based animal health systems are localised systems that provide an effective way for veterinary support to reach the remotest rural communities. They can also contribute to veterinary public health and livestock disease surveillance systems.

continued over

Technical standards
Veterinary support

LEGS Principle	Examples of how the principles are relevant in veterinary support interventions
7. Supporting local ownership *(continued)*	Local government and local NGOs should be involved in designing and monitoring interventions such as veterinary voucher schemes. Local government and NGOs should also be supported to design and deliver training courses (for example, for CAHWs) and to co-develop standards and guidelines for CAHW systems and voucher schemes.
8. Committing to MEAL	Monitoring, evaluation, accountability and learning (MEAL) is needed to track the progress of veterinary responses and make timely adjustments to implementation.
	More impact evaluation is needed to strengthen learning about emergency veterinary responses. There is very limited evidence on the livelihood impacts of emergency responses that support public sector veterinary functions.
	Monitoring systems and impact evaluations can provide valuable information to facilitate learning and improve future practice.

Protection

Veterinary workers carrying cash and/or high-value medicines may be vulnerable to robbery or attack. Insecurity can also have animal health implications. Animals stolen from a neighbouring group or area can introduce disease into the herd. In camps, the risk of livestock assets and associated goods being stolen is high *(see Chapter 7: Shelter)*.

Access

In remote areas with poor infrastructure and communications, veterinary service delivery is a challenge even in 'normal' times. Access to communities may only be achieved on foot or by boat. In camp-like settings, displaced livestock keepers may be beyond the reach of regular veterinary services. The more remote communities tend to be more at risk during an emergency. In these situations, veterinary paraprofessionals are usually the most appropriate service providers because they can travel to and function in these environments.

Technical standards
Veterinary support

Affordability and cost recovery

When providing veterinary support to communities, there are different approaches to cost recovery. Three options are outlined in *Box 6.1*.

Agencies responding to emergencies sometimes provide free veterinary support. This practice can threaten existing private services that depend on charging for veterinary care. It can also confuse livestock keepers, who receive services for a fee from some providers, then free from others. There is very limited evidence that providing free clinical veterinary care offers significant livelihood benefits to crisis-affected communities or is cost-effective or equitable. More evidence of livelihood benefits is available for veterinary paraprofessional systems that are based on some level of payment for services or which use voucher schemes.

Evidence shows poorer livestock keepers use private clinical services based on simple, low-cost, community-based approaches. However, during emergencies, agencies may face the challenge of providing rapid, equitable and effective clinical veterinary care while also supporting local private service providers who require an income. Approaches such as veterinary vouchers make clinical services available to poorer livestock owners while also supporting the private sector.

Box 6.1
Clinical veterinary service delivery in emergencies: Three options for payment

Services delivered free of charge: Coverage usually depends on the level of funding by external agencies. In many cases, services will reach only a small proportion of the crisis-affected community. If agency staff deliver clinical services, the likelihood of them undermining local services, markets and longer-term development processes is strong. Without supervision, there may also be a risk that services will not be provided free at the point of delivery.

Existing or newly trained veterinary paraprofessionals: Usually, these workers are paid by their community at rates lower than professional services would require. This approach helps to strengthen local capacity and supports systems that can be improved over time and as the emergency recedes. It also improves accessibility and availability. On the other hand, maintaining affordable services may be challenging.

> **Gradual introduction of payment for services:** In this option, services are provided free during the acute stage of an emergency. Providers then request payment for services in later stages as livestock markets begin to function. The risks of this option are similar to those of the first approach. It may be difficult for providers to persuade people that they need to pay if the service was previously provided free.

Use of cash and voucher assistance

During emergencies, veterinary professionals and paraprofessionals can partner with an agency to provide veterinary support. Delivery mechanisms, such as vouchers, can then be used to provide their services. *(See Chapter 3: Emergency response planning.)* Cash and voucher assistance can help reach poorer and more at-risk livestock users. It can also help to maintain private services during emergencies.

Agencies can provide cash and vouchers specifically for clinical veterinary services. Some public sector veterinary functions can be subsidised as a form of indirect grant. See *Process case studies: Veterinary voucher scheme in Kenya*, and *Veterinary voucher schemes in Ethiopia*, as well as *Impact case study: Vouchers for livestock distribution and veterinary support in Somalia*.

Camps

Camps with displaced people and their animals can create conditions for the spread of disease. For example, animals from different areas may have close contact or share feed or water troughs. Specific measures to reduce animal disease risk in camps should be considered. One way to do this is to establish quarantine areas where new arrivals are segregated from other animals for a period appropriate for the diseases of concern *(see Chapter 7: Shelter)*.

In camp settings, veterinary public health activities may be particularly appropriate. Participatory assessment of disease risks involves livestock keepers jointly identifying disease risks, including zoonotic diseases, and taking measures to prevent disease outbreaks.

Technical standards
Veterinary support

Decision tree for veterinary support options

The decision tree *(Figure 6.1)* summarises some of the key questions to consider in determining which may be the most feasible and appropriate option for an emergency veterinary support intervention. The standards, key actions and guidance notes that follow provide more information for detailed planning. Where possible, these should build on preparedness activities conducted prior to the onset of the emergency – in 'normal' times.

Technical standards
Veterinary support

Figure 6.1: Decision tree for veterinary support options

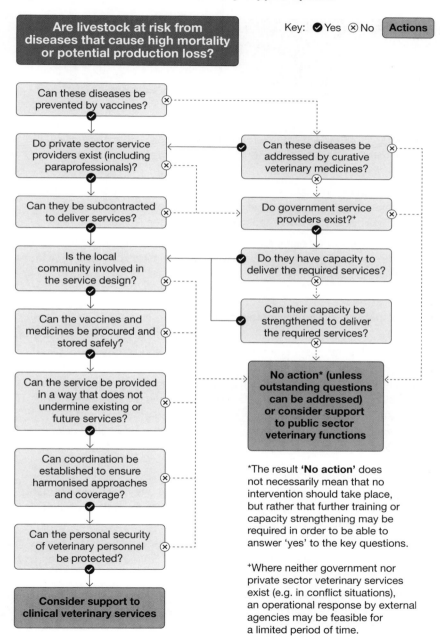

Technical standards
Veterinary support

The standards

Standard 1: Preparedness
Veterinary needs and capacities are assessed, and contingency plans are in place prior to the emergency.

Key actions

- Develop contingency plans for veterinary support with communities and local veterinary authorities and service providers *(see Guidance notes 1, 2, 3, 4 and 5)*.

- Ensure contingency plans are guided by an assessment of the quality of veterinary medicines from local suppliers, and of the capacity of local service providers to store medicines properly *(see Guidance note 6)*.

- Ensure contingency plans take account of the prevailing livestock disease situation specific to the local context and of the types of emergency in the area *(see Guidance note 7)*.

Guidance notes

1. Participatory contingency planning

In areas where livestock are an important livelihood asset and are at risk from rapid-onset or slow-onset emergencies, agencies should develop contingency plans for veterinary support. In line with LEGS *Principle 2: Community participation*, community members should help develop these plans, taking account of differences in the needs and priorities of different gender and wealth groups. Local veterinary service providers and government veterinary departments should also be involved in developing the contingency plans.

2. Policy and legal factors

Preparedness planning should include carrying out a review of government and agency policies, rules and procedures that relate to implementation options. For example:

- In some countries, certain types of veterinary paraprofessional workers are not legally recognised or are restricted to a limited range of veterinary activities.

- Some countries may have livestock disease control policies that need to be followed; if these are not followed, agencies need to justify alternative control methods.

- There may also be restrictions on using certain types of veterinary products, as defined by national drug registration bodies.
- The purchase of veterinary drugs is sometimes hindered by bureaucratic requirements from some donors and governments. These may prevent rapid and appropriate procurement in emergency contexts.
- Organisational or donor policy may hinder cost-recovery plans.

Understanding the policy context is vital, both for recognising potential constraints and as a foundation for future advocacy or policy action.

3. Flexible funding

Where possible, development programmes working in areas that are at risk of humanitarian crises should include flexible funding or crisis modifiers to enable early response. Clinical veterinary support is dynamic and relevant through all stages of emergencies. It should not be delayed or hindered by bureaucratic processes, and should be included as part of early response.

4. Service design

In cases where cash or vouchers are anticipated to access veterinary services, agencies can outline specific aspects of the scheme before an emergency occurs. These might include, for example, the monetary value of the voucher and the service fee required by the service provider. Model contracts and other administrative documents can be pre-prepared. For other aspects of service design to consider during preparedness, see *Standard 3: Clinical veterinary service design*.

5. Veterinary service providers

Given the importance of clinical veterinary care during emergencies, preparedness includes an understanding of the types of veterinary workers who are present in at-risk areas. Here, the emphasis should be on workers who have been trained to diagnose and treat locally occurring livestock diseases. Typically, these workers range from veterinarians to specific types of veterinary paraprofessionals, and they may be positioned in public or private sectors. Maintaining an inventory of such workers enables rapid responses. It also helps identify gaps in clinical capacities and related training or refresher training needs. Also see *Standard 2: Assessment and planning* and *Standard 4: Examination and treatment*.

Technical standards
Veterinary support

6. Medicine supply, quality and storage

Inventories of local veterinary service providers and suppliers of veterinary medicines should be developed and maintained. The quality of veterinary medicines is a critical issue. The quality of medicines from local suppliers should be assessed, and supply chains should be mapped. Visual assessment of medicines can be supported by laboratory testing of medicine quality. Medicine storage facilities should also be assessed (for example, the use of refrigeration). Suppliers of good-quality medicines with adequate storage facilities can be pre-positioned/selected. Local suppliers might also benefit from awareness raising or training on issues related to medicine quality and storage. In situations where veterinary medicines of suitable quality cannot be sourced locally, agencies should consider and plan for the importation of medicines. However, they should take into account the additional time required.

7. Assessment of livestock diseases

The livestock diseases that occur during an emergency may differ from those that occur during 'normal' periods. Participatory assessments carried out with livestock keepers can be used to identify and prioritise diseases that are specific to the main types of emergencies that affect the area. Participatory epidemiology methods are particularly useful for disease prioritisation. They can be supported with secondary information (for example, surveillance reports from local veterinary departments).

Standard 2: Assessment and planning

The crisis-affected population, including at-risk groups, actively participates in veterinary needs assessment and prioritisation.

Key actions

- Conduct rapid participatory veterinary needs assessment and prioritisation, involving all relevant subgroups within the crisis-affected population, and in partnership with local veterinary authorities and service providers *(see Guidance note 1)*.
- Within the affected area (or, for displaced communities, 'host community area'), map and analyse all existing veterinary service providers in terms of current and potential capacity if assisted by aid agencies *(see Guidance note 2)*.

Technical standards
Veterinary support

- Ensure the assessment before the emergency includes analysis of how pre-selected service providers prefer to be paid *(see Guidance note 2)*.
- Ensure the assessment includes a rapid analysis of policy or legal factors that may hinder or enable specific implementation strategies *(see Standard 1, Guidance note 3)*.

Guidance notes
1. Rapid participatory assessment
Rapid participatory assessment should:

- be conducted using experienced veterinary workers trained in participatory methods;
- include specific attention to the priorities of at-risk groups;
- involve consultation with local government and private sector veterinary personnel;
- aim to identify and prioritise livestock health problems warranting immediate attention according to livestock type and at-risk group;
- be cross-checked against secondary data of adequate quality where available.

A checklist and methods for assessment are given in *Appendix 6.1: Assessment methods and checklist for veterinary support*. (See also participatory data collection methods in *Chapter 3: Emergency response planning*.) Formal livestock disease surveys involving questionnaires and laboratory diagnosis are rarely feasible in emergency contexts. The modest added value of the disease information obtained is rarely justified in relation to the additional time and cost required, and the need for rapid action. During protracted crises, more systematic livestock disease surveys or studies may be necessary to refine disease control strategies. In these cases, participatory epidemiological approaches should be used as well.

2. Mapping and analysis of veterinary service providers
Drawing on contingency plans *(see Standard 1, Guidance note 5)*, agencies should develop a map of existing service providers (veterinary surgeons and all types of veterinary paraprofessional workers), their activities and coverage. This can be done rapidly if they have already developed an inventory of veterinary workers as part of preparedness *(see Standard 1: Preparedness)*. The map will assist agencies to define their strategy for service delivery, including planned geographical coverage and access to at-risk groups. They should review the pricing arrangements of the different

Technical standards
Veterinary support

service providers (which could also include traditional healers and other informal workers) as part of this mapping and analysis.

Categories of veterinary paraprofessional workers vary between countries but include:

- veterinary assistants;
- animal health auxiliaries/assistants;
- animal health technicians;
- CAHWs, as defined in national veterinary legislation and codes.

In some (usually conflict-based) emergencies, it is possible that neither the government nor the private sector can provide adequate veterinary services. In such cases, it may be appropriate for external agencies to support a community-based service through the training of CAHWs and/or livestock keepers. This should be based on plans for building government and/or private sector capacity as this becomes feasible as part of a clear exit strategy.

Standard 3: Clinical veterinary service design

Veterinary support is designed appropriately for the local social, technical, security and policy context with the active participation of crisis-affected communities.

Key actions

- Ensure the service design process uses the information and analyses of the contingency plan and/or the initial assessment. Also ensure it is based on the active participation of the crisis-affected population, including at-risk groups *(see Guidance note 1)*.
- Check that the service design includes specific elements to reach at-risk groups and, in particular, addresses challenges of accessibility and affordability *(see Guidance note 2)*.
- Ensure that the service design considers the need for rapid procurement and availability of relevant veterinary vaccines and medicines. Also ensure it considers the need for appropriate quality of products and proper storage at field level *(see Guidance note 3)*.
- Check that the service design includes plans for rapid training of local service providers as necessary *(see Guidance note 4)*.

Technical standards
Veterinary support

- Ensure that the service design is based on local social and cultural norms, particularly in relation to gender roles *(see Guidance note 5)*.
- Ensure that the service design maximises the security of local people, veterinary service providers and aid agency staff *(see Guidance note 6)*.
- Ensure that the service design incorporates payment for services, where possible *(see Guidance note 7)*.
- Ensure that the service design builds in the professional supervision of veterinary paraprofessionals *(see Guidance note 8)*.

Guidance notes

1. Design based on assessment findings

Service design should aim to address the prioritised livestock health problems identified in the contingency plan and verified during the initial assessment *(see Chapter 3: Emergency response planning)*. It is rarely feasible or appropriate for a primary-level veterinary service to address all livestock health problems. In most cases, a limited range of vaccines and medicines can prevent or treat the most important diseases in a given area.

The focus of the service on prioritised livestock diseases needs to be understood and agreed upon by all actors, including livestock keepers. Where the service cannot address the priority (for instance, when necessary facilities such as a cold chain are unavailable), this should be agreed upon with all stakeholders, including the affected communities. Similarly, appropriate timing for interventions (particularly vaccination) should be discussed and agreed upon with all stakeholders. The affected communities should be actively involved in the design of the service as far as possible.

2. Reaching at-risk groups

Service design should consider the types of livestock that at-risk groups own or use, and should address the health problems of these types of livestock. Special attention should be given to accessibility and affordability issues in order to promote equitable access. Access to remote areas with limited infrastructure may require expensive means of transport (by air, for example), which limits coverage within a given budget. Alternatively, access can be achieved by using locally based veterinary paraprofessional workers, who can travel on foot, mules, bicycles, boats or by other local means of transport. In some cases, programmes may need to provide or support local modes of transport for veterinary workers.

In rapid-onset emergencies, agencies might provide transport free of charge. In more protracted crises, cost-sharing arrangements are often feasible. The payment-for-services strategy needs to take account of the need for rapid and equitable delivery, while also supporting private sector veterinary workers where possible. For more at-risk groups, private veterinary workers can be subcontracted to deliver a service for a specified short period of time. They may use voucher schemes *(see Process case studies: Veterinary voucher scheme in Kenya and Veterinary voucher schemes in Ethiopia)*. In areas where the private veterinary sector is active or where the government charges for clinical veterinary care, normal pricing policies should be followed. These circumstances can involve possible exemptions for targeted at-risk groups. To avoid confusion, community participation and the agreement of community representatives on these issues will be needed, as well as clear communication with all stakeholders.

3. Rapid procurement and storage

Agencies with limited experience in veterinary drug procurement should seek expert advice. The quality of veterinary drugs and vaccines varies considerably between suppliers, whether sourced locally or internationally. Suppliers vary in their capacity to supply medicines in large volumes with appropriate expiry dates within agreed delivery times. The wide range of products available can further complicate procurement. Because some veterinary vaccines require isolation of local field strains of disease pathogens, agencies need to verify the vaccine's exact composition. Local importers, often located in capital cities, can supply readily available drugs in reasonable quantities. However, the quality, expiry dates and drug storage conditions should be checked. At the local level, many veterinary vaccines and some drugs require cold storage. They should not be purchased or used unless adequate cold storage facilities are in place, as well as a cold chain for transporting and storing them. Storage in camp-like settings may present particular challenges because of the lack of cold chain maintenance and storage. Cold storage facilities for human health services can sometimes be shared. However, human health professionals are sometimes unwilling to store veterinary products in human health cold chains. A high-level agreement needs to be reached beforehand to take full advantage of expensive cold chain facilities.

4. Community-based approaches

Where some veterinary workers are already present, and rapid delivery of services is required, training should be limited to short refresher courses. These should focus on 1) clinical diagnosis of the prioritised diseases and 2) correct use of veterinary vaccines or drugs. Depending on the existing capacity of local personnel, this refresher training is not always needed. Where agencies need to select and train veterinary paraprofessional workers such as CAHWs from scratch, guidelines are available for CAHW systems in development programmes rather than emergency programmes. These guidelines note the importance of linking CAHWs to higher-level paraprofessionals or veterinarians to support medicine supply to CAHWs and for CAHW supervision (also see *Guidance note 8*). To enable rapid response in emergency situations, agencies may need to streamline some good practice principles relating to CAHW selection and training. However, as emergencies become protracted or come to an end, further training is recommended to enhance CAHW knowledge and skills. In some countries, national technical intervention standards and guidelines for CAHW systems are available, as well as training manuals for short, practical, participatory CAHW training courses.

5. Social and cultural norms

The design of veterinary support needs to take account of local social and cultural norms, particularly those relating to the roles of men and women as service providers. In some communities, it is difficult for women to handle some livestock species, move freely or travel alone to more remote areas where livestock might be present. However, even in very conservative cultures, it is often possible for women to be selected and trained by women as CAHWs to provide the service to other women.

6. Protection

Where livestock are very important to local economies and livelihoods, veterinary drugs are highly prized. These small-volume, high-value items are easy to steal and resell. Service design should consider the risk to veterinary personnel of violence, abduction or theft. Livestock are often grazed away from more secure settlements. Sometimes they are moved long distances to grazing areas and water points. Veterinary workers travelling to such areas may be at risk, especially in conflict situations. Local veterinary paraprofessional workers may be appropriate in these situations. This is because they know the area and may be familiar enough with armed groups or security forces to be able to negotiate access.

7. Payment for service

Based on evidence, service design should incorporate payment for services where possible. Voucher schemes should be used for the most at-risk livestock keepers. Other livestock keepers should rapidly resume full payment for services. Governments may consider all vaccination as a 'public good' rather than a 'private good'. However, prevention of diseases not easily transmitted between animals, such as clostridial diseases, may be considered as a private good. Theoretically, the private sector is best equipped to deliver private goods.

8. Professional supervision of veterinary paraprofessional workers

Even where paraprofessionals such as CAHWs are working in remote areas, they should be under the overall supervision of a veterinarian. Professional supervision enables monitoring of the correct use of veterinary products, disease reporting to the authorities, and integration of CAHWs into existing private veterinary services. In some situations, the use of mobile phones can improve supervision of CAHWs.

> **Standard 4: Examination and treatment**
> Clinically trained veterinary workers conduct examination and treatment with the active participation of the affected communities.

Key actions

- Maximise the use of veterinarians and clinically trained veterinary paraprofessionals for the diagnosis and treatment of livestock diseases *(see Guidance note 1)*.
- Clearly document the roles and responsibilities of all actors. Where appropriate and necessary, make written agreements *(see Guidance note 2)*.
- Euthanise incurable sick or injured animals humanely and safely *(see Guidance note 3)*.

Guidance notes

1. Use of veterinarians and clinically trained veterinary paraprofessionals

In many humanitarian situations, livestock diseases will be recognised by livestock keepers. Indigenous knowledge on livestock diseases is an important resource to assist diagnosis. In some situations, physical access to animals by trained veterinary workers will be limited or impossible, and therefore veterinary staff depend on reports of diseases provided by livestock keepers. When veterinary staff can access livestock, the diagnosis will depend heavily on the clinical examination because laboratory support is often not available. In these situations, the clinical skills and knowledge of veterinarians and clinically trained veterinary paraprofessionals should be used for diagnosis and treatment. Their clinical training will likely be based on curricula in local veterinary schools and training institutes. These should cover the livestock species and diseases of relevance locally.

In complex emergencies, agencies should consider some support to local veterinary laboratories to assist disease diagnosis or establish a basic laboratory. The correct diagnosis of disease with laboratory support is especially important in the case of certain epidemic diseases and where large-scale vaccination is being considered. Misdiagnosis in these situations would lead to incorrect vaccines being used. Laboratory diagnosis can also support disease surveillance *(see Standard 8: Livestock disease surveillance)*.

2. Roles and responsibilities

During emergency clinical veterinary service provision, problems may occur due to lack of stakeholder coordination. For example, problems can arise from a misunderstanding of the roles and responsibilities of different actors. These include false expectations about the service's aims and coverage, or confusion over pricing arrangements or recipient selection. Many of these problems can be avoided by being committed to community participation and consulting stakeholders. Where possible, there should be close collaboration with local authorities and private sector actors. Roles and responsibilities should be documented in memoranda of understanding or similar agreements. These can provide useful points of reference in subsequent disputes.

3. Euthanasia

Animal euthanasia should follow humane standards and practices. Depending on the sickness/injury and method of slaughter, some livestock carcasses may be fit for human consumption *(see Standard 6: Sanitation and food hygiene)*. Criteria for euthanasia should follow international or government guidelines, as should the procedures for euthanising and safe carcass disposal. Religious and traditional considerations with regard to slaughter also need to be taken into account.

Technical standards
Veterinary support

> ## Standard 5: Zoonotic diseases
> The crisis-affected population has access to information and services designed to prevent and control zoonotic diseases.

Key actions

- Include participatory assessment of zoonotic diseases and their prioritisation in the initial assessment of animal health problems *(see Guidance note 1)*.

- Design and implement zoonotic disease control measures either in conjunction with the provision of clinical services or as a stand-alone activity *(see Guidance note 2)*.

Guidance notes

1. Assessment

The contingency plan *(Standard 1: Preparedness)* and/or rapid participatory assessment *(conducted under Standard 3: Clinical veterinary service design)* should include assessment of zoonotic diseases in terms of actual cases or risk. During emergencies, zoonotic disease risk may increase or decrease. Examples include 1) anthrax associated with abnormal movement of livestock to grazing areas that are normally avoided; 2) rabies associated with local populations of wild or domestic predators, possibly attracted to carcasses or garbage; 3) zoonotic disease associated with close contact between animals and people; 4) unhygienic conditions arising from the crowding of people and animals in camps; and 5) water supply breakdown.

2. Zoonotic disease control

The disease control method varies according to the zoonotic diseases in question. For some diseases, veterinary paraprofessionals may provide information to livestock keepers verbally or by using leaflets. Such workers might also assist with organising vaccination campaigns, for example, against rabies, or with the humane control of stray dog populations. Outreach to women can be particularly important because women can play a significant role in livestock health management. Yet they are often overlooked in disease control measures. Where private workers are used on a short-term basis, payment for their services by an external agency is usually required. Zoonotic disease control efforts should be harmonised between agencies and between areas as part of the coordination effort. Collaboration with human health agencies and programmes helps harmonise

Technical standards
Veterinary support

approaches and enables the sharing of resources such as cold-storage facilities *(see Standard 3, Guidance note 3)*.

> ## Standard 6: Sanitation and food hygiene
> Sanitary and food hygiene measures relating to the consumption of livestock products and the disposal of livestock are established.

Key actions

- Construct slaughter slabs during protracted crises *(see Guidance note 1)*.
- Establish meat inspection procedures at slaughter slabs and abattoirs used by the affected population *(see Guidance note 1)*.
- Publicise good food-handling practices *(see Guidance note 2)*.

Guidance notes

1. Slaughter facilities and meat inspection

In camp-like settings or in situations in which slaughter facilities have been damaged, it may be appropriate to construct slaughter slabs. This will encourage humane slaughter as well as hygienic handling and inspection by trained workers. Similarly, animal welfare, health, and hygiene standards will need to be met for livestock offtake for slaughter. Here, either fixed or mobile slaughter slabs may need to be constructed *(see Chapter 8: Livestock offtake)*. In all these cases, consultation with local livestock workers or butchers will help to determine the correct locations for slaughter slabs and their design. Meat inspection procedures are generally well known to animal health personnel. Safe disposal of offal from slaughtered livestock should be ensured.

2. Public awareness

Based on the findings of the assessment, public education campaigns should be conducted as appropriate to raise awareness of best practices in safe food handling and preparation. For example, campaigns can give advice to control tuberculosis or brucellosis through improved hygiene when handling either animals or meat, or when preparing food, and by encouraging consumption of boiled milk.

Technical standards
Veterinary support

Standard 7: Disposal of dead animals
Dead animal disposal is organised hygienically according to need.

Key actions
- Assess the needs for disposal *(see Guidance note 1)*.
- Dispose of carcasses to ensure good hygiene *(see Guidance note 2)*.

Guidance notes
1. Disposal needs assessment
When natural hazards such as fire or earthquakes occur, many animals may be injured, and either treatment or euthanasia may be required. Slow-onset emergencies such as drought and severe winter may cause large numbers of animal deaths, as may widespread floods or cyclones. The hygienic disposal of animal carcasses then needs to be considered. Animal carcasses may spread disease, are unsightly, produce noxious odours, and attract scavengers such as packs of dogs, hyenas or jackals, and crows and vultures. In droughts and winter emergencies, animals die mainly from undernutrition, dehydration and hypothermia respectively, and not from diseases. Yet disease agents may remain in carcasses and pose risks to human and animal health. Another key consideration may be the psychological effect on livestock keepers of constantly seeing their dead animals. On these grounds alone, it may be justifiable to organise disposal.

2. Disposal
Environmental and health considerations should be taken into account. Composting can be an effective way to dispose of animal bodies that also produces useful fertiliser. Burying animals anywhere where water sources may be contaminated should be avoided. Cash for work schemes, in which community members are paid to undertake carcass disposal, have been used effectively *(see Process case study: Carcass disposal in Mongolia)*. See *FAO (2016)* for technical details on carcass disposal, including composting.

Technical standards
Veterinary support

Standard 8: Livestock disease surveillance

During protracted emergencies, a livestock disease surveillance system is supported that ensures timely response to disease outbreaks.

Key actions

- Conduct livestock disease investigation in response to disease outbreaks to confirm diagnosis and trace the source of disease as well as where it may have spread. This can also instigate or modify control measures as necessary *(see Guidance note 1)*.
- In protracted crises, and for livestock diseases covered by national disease surveillance policies or eradication strategies, collect information in line with these policies and strategies *(see Guidance note 2)*.
- Ensure the coordination body compiles livestock disease data and submits the compiled report to the relevant veterinary authority *(see Guidance note 3)*.

Guidance notes

1. Veterinary investigation and response

Veterinary initiatives and agencies should have the capacity to conduct investigations of disease outbreaks. Within multi-agency initiatives, this task may be entrusted to a team or individual with specialist training in disease investigation, including post-mortem examination and laboratory diagnosis. In the absence of such assistance, agencies should be prepared to collect relevant samples and submit them to a diagnostic laboratory, either in country or abroad. All activities need to complement government veterinary investigation systems, where these exist, with official reporting of diagnoses being made by government actors. During protracted crises, agencies should consider establishing a small, local diagnostic laboratory to support the capacity of clinical veterinary workers and disease investigations. Sharing facilities with medical laboratories may be feasible. Investigators should use standard recording forms with checklist questions to assist with collecting relevant information for tracing disease sources and spread. To ensure the continued support of communities, veterinary investigation needs to be closely linked to timely responses to control a disease outbreak (for example, through the treatment or vaccination of livestock).

Technical standards
Veterinary support

2. Animal disease surveillance

In many countries, specific animal diseases have national or international control or eradication programmes. Standardised surveillance procedures are set by international organisations such as WOAH and FAO. Where possible, livestock disease surveillance systems in protracted crises should follow these procedures. Operational constraints may prevent the implementation of standard surveillance procedures. Here, liaison with national authorities (if working) and either WOAH or FAO can enable modified surveillance methods to suit the conditions.

3. Reporting

In protracted crises, all agencies should submit regular (usually monthly) surveillance reports to the coordination body for compilation and submission to the relevant government authority. They should provide brief reports that summarise pooled surveillance data from the region to veterinary workers who submit data.

Technical standards
Veterinary support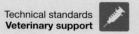

Appendices

Appendix 6.1: Assessment methods and checklist for veterinary support

Indicator	Useful method*
1. Accessibility: The physical distance between livestock keepers and the nearest trained veterinary workers	**Participatory mapping:** Simple sketch maps of a given area with: – locations and owners of livestock – nearest veterinary services/types – distance (km, hours, etc.)
2. Availability: A measure of a service's physical presence and concentration/availability in an area	**Participatory mapping:** As above **Direct observation:** Veterinary workers Facilities **Interviews:** Assess existing stocks of veterinary products Quality of medicines and equipment Barriers to availability based on caste, ethnicity, gender, etc.
3. Affordability: The ability of people to pay for services	**Semi-structured interviews** **Observation:** Veterinary facilities Livestock markets Price lists (These will determine normal service costs and livestock values, and allow comparison of service costs against livestock worth. If livestock markets are still functioning, or if livestock offtake is taking place, it is more likely that people will be able to pay for veterinary support. See also MISMA.)
4. Acceptance: Relates to cultural and political acceptance of veterinary workers, which is affected by sociocultural norms, gender issues, language capabilities, and other issues	**Interviews:** with male and female livestock keepers (young and old)

continued over

Technical standards
Veterinary support

Indicator	Useful method*
5. Quality: This includes veterinary workers': level of training technical knowledge and skills communication skills and the quality and range of veterinary medicines and vaccines, or access to equipment	**Interviews:** Veterinary workers **Direct observation:** Veterinary facilities Education certificates Licences to practise or equivalent
All indicators	**Matrix scoring:** Scoring different types of veterinary workers operational in the area against the five indicators shows the relative strengths and weaknesses of each type.

See Suggested participatory methods for carrying out initial assessment in Chapter 3: Emergency response planning.

Appendix 6.2: Examples of monitoring and evaluation indicators for veterinary support

	Process indicators (measure things happening)	Impact indicators (measure the result of things happening)
Designing the intervention	Completion of participatory survey and analysis	Identification of most important animal health problems in the community according to different wealth and gender groups
	Number of meetings with community/community representatives	Analysis of options for improving animal health
	Number of meetings between private veterinary workers and implementing agency	**Veterinary vouchers:** Value of vouchers agreed with community and local private veterinary service providers
		Affected community selection criteria agreed
		Number of veterinary paraprofessionals linked to private veterinary drug supplier or agency
		Reimbursement system for private sector workers and suppliers agreed
		Monitoring system agreed
		Implementing agency provides medicines: Number of veterinary paraprofessionals supplied by agency and geographical coverage

continued over

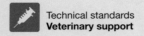

Technical standards
Veterinary support

	Process indicators (measure things happening)	Impact indicators (measure the result of things happening)
Rapid veterinary training/refresher training	Number and gender of workers trained Number and type of animal health problems covered in training course Cost of training	Improved veterinary knowledge and skills among trainees
Veterinary activities	**Veterinary vouchers:** Number of vouchers distributed by area and type of household Number of treatments per disease per livestock type per household Number and value of vouchers reimbursed **Medicines provided by agency or private veterinary pharmacy:** Quantities and types of medicines supplied to veterinary workers Cost of medicines supplied to veterinary workers Number of treatments per disease per livestock type per worker per month Number of monitoring forms submitted by veterinary workers Number of disease outbreaks reported by veterinary workers	Livestock mortality by species and disease against baseline Geographical coverage of veterinary workers Proportion of livestock-rearing households serviced Proportion or number of workers functioning after training Action taken according to disease outbreak reports Human nutrition – consumption of animal-sourced foods in community in relation to improved animal health and according to wealth and gender groups Income in community in relation to improved animal health and according to wealth and gender groups Influence on policy

References and further reading

Admassu, B., Nega, S., Haile, T., Abera, B., Hussein A., Catley, A. (2005) 'Impact assessment of a community-based animal health project in Dollo Ado and Dollo Bay districts, southern Ethiopia', *Tropical Animal Health and Production* 37: 33–48, http://dx.doi.org/10.1023/B:TROP.0000047932.70025.44

Aklilu, Y. (2003) 'The impact of relief aid on community-based animal health programmes: the Kenyan experience', in K. Sones and A. Catley (eds), *Primary Animal Health Care in the 21st Century: Shaping the Rules, Policies and Institutions*, proceedings of an international conference, 15–18 October 2002, Mombasa, Kenya, African Union/Interafrican Bureau for Animal Resources, Nairobi, http://www.eldis.org/go/home&id=13563&type=Document#.U38tmbkU_Vh

Baker, J. (2011) *5-year evaluation of the Central Emergency Response Fund, country study: Mongolia*, Channel Research, Belgium on behalf of the United Nations Office for the Coordination of Humanitarian Affairs (OCHA), http://www.alnap.org/resource/11664

Bekele, G., Akumu, J., (2009) *Impact assessment of the community animal health system in Mandera West, Kenya*, Tufts University, CARE, Save the Children, VSF Suisse, Nairobi, https://fic.tufts.edu/pacaps-project/Coordination%20Support/VSF%20ELMT%20CAH%20PIA%20report.pdf

Braam D.H., Chandio R., Jephcott F.L., Tasker A., Wood J.L.N. (2021) 'Disaster displacement and zoonotic disease dynamics: The impact of structural and chronic drivers in Sindh, Pakistan', *PLOS Glob Public Health* 1(12): e0000068, https://doi.org/10.1371/journal.pgph.0000068

Burton, R. (2018) *Humane destruction of stock*, PrimeFact 310, Fourth edition, New South Wales Department of Primary Industries, Orange, New South Wales, https://www.dpi.nsw.gov.au/__data/assets/pdf_file/0007/1309489/Humane-destruction-of-stock.pdf

Catley, A., Admassu, B., Bekele, G. and Abebe, D. (2014) 'Livestock mortality in pastoralist herds in Ethiopia during drought and implications for drought response' *Disasters* 38(3), 500-516, https://doi.org/10.1111/disa.12060

Catley, A., Leyland, T. and Blakeway, S. (eds) (2002) *Community-based Animal Healthcare: A Practical Guide to Improving Primary Veterinary Services*, ITDG Publishing, London.

Catley, A., Leyland, T., Mariner, J.C., Akabwai, D.M.O., Admassu, B., Asfaw, W., Bekele, G., and Hassan, H.Sh., (2004) 'Para-veterinary professionals and the development of quality, self-sustaining community-based services', *Revue scientifique et technique de l'office international des épizooties* 23(1): 225–252, http://www.livestock-emergency.net/userfiles/file/veterinary-services/Catley-et-al-2004.pdf

Catley, A. (2020) 'Participatory epidemiology: reviewing experiences with contexts and actions', *Preventive Veterinary Medicine* 180, 105026, https://doi.org/10.1016/j.prevetmed.2020.105026

Catley, A. (2005) *Participatory Epidemiology: A Guide for Trainers*, African Union/Interafrican Bureau for Animal Resources, Nairobi.

Catley, A., Abebe, D., Admassu, B., Bekele, G., Abera, B., Eshete, G., Rufael, T. and Haile, T. (2009) 'Impact of drought-related livestock vaccination in pastoralist areas of Ethiopia', *Disasters* 33(4): 665–685, http://dx.doi.org/10.1111/j.1467-7717.2009.01103.x

FAO (2011) *The use of cash transfers in livestock emergencies and their incorporation into Livestock Emergency Guidelines and Standards (LEGS)*. Animal Production and Health Working Paper. No. 1. Rome, https://www.fao.org/3/i2256e/i2256e00.pdf

Technical standards
Veterinary support

FAO (2012) *An assessment of the impact of emergency de-worming activities*, FAO report OSRO/KEN/104/EC – FAOR/LOA NO. 008/2012, FAO, Nairobi. https://www.livestock-emergency.net/wp-content/uploads/2020/05/RVC-WSPA-FAO-ILRI-2012.pdf

FAO (2016) *Livestock-related interventions during emergencies – The how-to-do-it manual*. Edited by Philippe Ankers, Suzan Bishop, Simon Mack and Klaas Dietze. FAO Animal Production and Health Manual No. 18. Rome, https://www.fao.org/3/i5904e/i5904e.pdf(See Chapter 5)

Farm Africa (2006) *Immediate support to agro-pastoral communities as a drought mitigation response: Marsabit and Moyale Districts*, Final Report to FAO, OSRO/RAF/608/NET (CERF2), Farm Africa, Nairobi.

Gary, F., Clauss, M., Bonbon, E., Myers, L (2021) *Good emergency management practice: The essentials – A guide to preparing for animal health emergencies*, 3rd edn, FAO Animal Production and Health Manual No. 25. Rome, FAO, https://doi.org/10.4060/cb3833en

Heath, S.E., Kenyon, S.J., and Zepeda Sein, C.A. (1999) 'Emergency management of disasters involving livestock in developing countries', *Revue scientifique et technique de l'office international des épizooties* 18(1): 256–271, https://doi.org/10.20506/rst.18.1.1158

Hufnagel, H. (2020) *The quality of veterinary pharmaceuticals: a discussion paper for the Livestock Emergency Guidelines and Standards*, LEGS, https://www.livestock-emergency.net/wp-content/uploads/2020/11/LEGS-Discussion-Paper-The-Quality-of-Veterinary-Pharmaceuticals-1.pdf

Iles, K. (2002) 'Participative training approaches and methods' and 'How to design and implement training courses', in A. Catley, T. Leyland, and S. Blakeway (eds), *Community-based Animal Healthcare: A Practical Guide to Improving Primary Veterinary Services*, ITDG Publishing, Rugby.

LEGS (2020) *The challenges of emergency veterinary voucher schemes: research into operational barriers to applying LEGS*, LEGS Briefing Paper. https://www.livestock-emergency.net/wp-content/uploads/2020/02/The-Challenges-of-Emergency-Veterinary-Voucher-Schemes-LEGS-Briefing-Paper-January-2020.pdf

LEGS (2020) *LEGS core standards and community-based animal health services*, LEGS Briefing Paper, https://www.livestock-emergency.net/wp-content/uploads/2020/02/LEGS-Core-Standards-and-Community-Based-Animal-Health-Services-LEGS-Briefing-Paper-January-2020.pdf

Leyland, T. (1996) 'The case for a community-based approach, with reference to Southern Sudan', in *The World Without Rinderpest*, pp. 109–120, FAO Animal Health and Production Paper 129, FAO, Rome.

Leyland, T., Lotira, R., Abebe, D., Bekele, G. and Catley, A. (2014) *Community-based animal health care in the Horn of Africa: an evaluation for the US Office for Foreign Disaster Assistance*, Feinstein International Center, Tufts University, Addis Ababa and Vetwork UK, Great Holland.

Linnabary, R.D., New, J.C. and Casper, J. (1993) 'Environmental disasters and veterinarians' response', *Journal of the American Veterinary Medical Association* 202(7): 1091–1093.

Mutungi, P.M. (2005) *External evaluation of the ICRC veterinary vouchers system for emergency intervention in Turkana and West Pokot districts*, International Committee of the Red Cross (ICRC), Nairobi.

Okell, C.N., Mariner, J., Allport, R., Buono, N., Mutembei, H.M., Rushton, J., Verheyen, K. (2016) 'Anthelmintic administration to small ruminants in emergency drought responses: assessing the impact in two locations of northern Kenya', *Tropical Animal Health and Production* 48:493–500, https://www.livestock-emergency.net/wp-content/uploads/2020/05/Okell-at-al-2016.pdf

Regassa, G. and Tola, T. (2010) *Livestock emergency responses: the case of treatment voucher schemes in Ethiopia*, FAO, Addis Ababa.

Schreuder, B. (2015) *Afghanistan, a 25-years' Struggle for a Better Life for its People and Livestock*, DCA-VET/Erasmus Publishing, Lelystad/Rotterdam, https://dca-livestock.org/books/

Schreuder, B.E.C., Moll, H.A.J., Noorman, N., Halimi, M., Kroese, A.H., and Wassink, G. (1995) 'A benefit-cost analysis of veterinary interventions in Afghanistan based on a livestock mortality study', *Preventive Veterinary Medicine* 26: 303–314, http://dx.doi.org/10.1016/0167-5877(95)00542-0

Simachew, K. (2009) *Veterinary voucher schemes: an emergency livestock health intervention – case studies from Somali Regional State, Ethiopia*, Save the Children USA, Addis Ababa.

UNDP (United Nations Development Programme) (2010) *2010 Dzud Early Recovery Programme*, UNDP Project Document, United Nations Development Programme, Ulaanbaatar.

Vetwork UK (2019) *Operational barriers to applying LEGS: research report*, https://www.livestock-emergency.net/wp-content/uploads/2020/02/LEGS-Research-Report-Operational-Barriers-to-Applying-LEGS-December-2019.pdf

World Health Organization (2021) 'Tripartite and UNEP support OHHLEP's definition of "One Health"', https://www.who.int/news/item/01-12-2021-tripartite-and-unep-support-ohhlep-s-definition-of-one-health

See also case studies for veterinary support interventions in emergencies at: https://www.livestock-emergency.net/resources/case-studies/

Technical standards
Livestock shelter and settlement

Chapter 7: Technical standards for livestock shelter and settlement

251   **Introduction**

254   **Options for shelter and settlement**

259   **Timing of interventions**

260   **Links to other LEGS chapters and other HSP standards**

261   **LEGS Principles and other issues to consider**

265   **Decision tree for livestock shelter and settlement options**

268   **The standards**

282   **Appendix 7.1: Assessment checklist for livestock shelter and settlement provision**

284   **Appendix 7.2: Examples of monitoring and evaluation indicators for livestock shelter and settlement**

285   **Appendix 7.3: Livestock shelter and climate challenges**

287   **References and further reading**

Photo © The Brooke

Technical standards
Livestock shelter and settlement

Chapter 7: Technical standards for livestock shelter and settlement

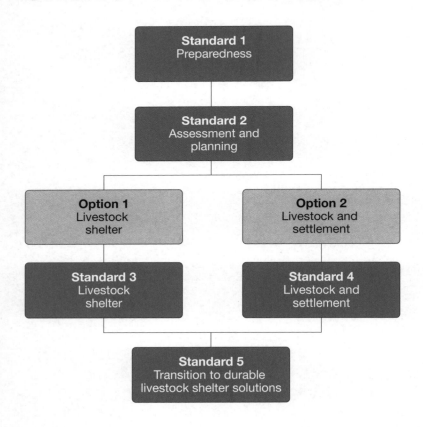

Technical standards
Livestock shelter and settlement

Introduction

Livestock shelters are the physical structures that, in some environments, animals need to survive. Not all livestock require shelter, or their need for shelter might be seasonal. Shelters can be either temporary or longer-lasting, according to need. Needs include **protection** against cold/hot climates, predators, and/or theft; **prevention** from wandering; and provision of a **healthy environment** for livestock and people. Shelters are also often used to manage livestock or as an animal husbandry practice.

Livestock settlement is the term used to refer to the wider context of livestock shelters. It includes land rights, environmental implications of shelter provision and access to feed and water. During emergencies, there are many areas where shelter and settlement needs for people and livestock overlap. Shelter and settlement for people is covered in detail in the Sphere Handbook.

Whether a community is displaced or non-displaced by an emergency is a key determinant of how they recover their shelter and livelihoods. When communities have been displaced, they may arrive in areas that are significantly hotter, colder, wetter or drier than they have been used to. They may also have greater need for livestock shelter. During complex emergencies, security often becomes a significant shelter and settlement issue.

This chapter presents information on supporting livestock shelter and settlement as an emergency livestock response, together with the technical options for each, and their associated benefits and challenges. Information is also available in Chapter 8, 'Livestock shelter and settlement' in *FAO* (2016). For each technical option, LEGS provides information through standards, key actions, and guidance notes. Checklists for assessment, as well as for monitoring and evaluation indicators, are presented in the appendices at the end of this chapter, and further reading is provided too. Case studies are presented on the LEGS website (see https://www.livestock-emergency.net/resources/case-studies/).

Links to the LEGS livelihoods objectives

Livestock shelter and settlement relates primarily to the second of the LEGS livelihoods objectives: to support crisis-affected communities to **protect key livestock assets**.

**Technical standards
Livestock shelter and settlement**

Livestock shelter can be vital to ensure livestock survive an emergency and continue to support livelihoods. As well as designing shelters to protect livestock assets after an emergency, it is important to include emergency mitigation measures (for example, earthquake-resistant livestock shelters) as preparedness before an emergency. 'Build-back-safer' elements are also important for the design of long-term shelters.

The importance of livestock shelter and settlement in emergency response

Livestock shelter interventions may be appropriate before, during and following an emergency. They may either replace the structures for previously sheltered animals, or construct new livestock shelters in response to a new context. Some examples of interventions are:

- when previously sheltered animals lose their shelter, for example, as a result of a flood or earthquake in which structures have been destroyed;
- when livestock keepers are displaced because of an emergency, and their livestock lose access to previous shelter, or the context requires new shelter (for example, when they move into camps);
- when extreme weather conditions (heat or cold) or conflict and insecurity require new shelter for previously unsheltered livestock;
- when livestock have been distributed as part of the response, and new shelters are required to protect them from weather, theft or predators.

The provision of livestock shelter as part of an emergency response also contributes to three of the animal welfare 'domains' described in *Chapter 1: Introduction to LEGS*, namely 'Environment', 'Health' and 'Mental state'.

The safety, security and welfare of their livestock are a primary concern of livestock keepers during or following all types of emergencies. There are many cases of livestock keepers prioritising the shelter needs of their livestock. This is irrespective of whether or not support is provided by intervening agencies, and in some cases in preference to safe shelter for their own family members *(see Box 7.1).*

Technical standards
Livestock shelter and settlement

Box 7.1
Actions of livestock keepers in prioritising shelter

In areas prone to yearly flooding, communities may construct flat-topped and compacted earth mounds, or else platforms on wooden pillars. Animals can then be herded onto these in response to flood warnings. There may not be enough space for both livestock and humans, or livestock protection may be too far away from safe areas for humans. In such cases, livestock keepers may forego their own shelter to remain close to their livestock.

In climates with cold seasons, displaced families may cohabit with their animals in shared shelters. The families benefit from the body heat of the livestock during the winter nights, although co-location increases the potential for zoonotic disease transmission.

In complex emergencies, co-location with animals may help to reduce the risk of livestock being stolen. If there is not sufficient safe space for livestock within camps, displaced livestock keepers may prefer to remain outside the camps with their animals. They might do this even if it increases their general insecurity, and results in less access to basic services in the camp, such as health or education.

Displaced livestock keepers sometimes use materials distributed for their own shelter to make shelters for their livestock. They may do so even if this compromises the effectiveness or safety of the shelter intended for human use. In Pakistan after the 2005 earthquake, shelter agencies distributed kits of shelter materials, including nylon rope intended to bind the wooden joints of the shelters together. Many families chose instead to use the nylon rope to tether or make enclosures for their goats. Shelter experts then worked with the communities to identify scrap metal from the rubble of the destroyed houses to use instead of the nylon rope to secure the wooden joints of the shelters. This meant the families could have an earthquake-resistant shelter and look after their livestock at the same time.

(Source: J. Kennedy, *personal comment*.)

Technical standards
Livestock shelter and settlement

Options for shelter and settlement

Despite evidence of the importance of livestock shelter to livestock keepers, it is an underdeveloped component of emergency response. There are limited examples of effective interventions in this area. In line with LEGS *Principle 2: Community participation*, crisis-affected communities should be supported in choosing their own shelter options. These should be based on how they judge their needs and on their preferred options for recovery. Agencies need to recognise these priorities, providing sufficient shelter support for livestock as well as for people. Other key elements to consider are whether the affected community has been displaced or not, and the climate.

For people, the options for shelter and settlement support have become wide-ranging. They include alternatives to simply providing physical shelters or shelter construction materials. Flexible support options allow more efficient use of limited resources in times of crisis. Cash and voucher assistance (CVA) and market-based approaches, together with support for housing, land and property rights (HLP), have become central to many human shelter responses. They provide greater dignity and decision-making power to emergency-affected households. These different shelter options should also be used to allow livestock keepers to define their shelter priorities for their livestock.

In terms of human shelter, **camps** are described as an option of last resort: to be avoided if other better options are available. Other options may provide more safety and dignity to households, as well as closer access to livelihood opportunities and basic services such as health and education. For major assets such as livestock, however, a camp may be the preferred option as it can provide sufficient space and security.

Livestock shelter and settlements for displaced and non-displaced populations

People's physical displacement during an emergency obstructs their options for recovery. Displacement over an international border may affect whether local authorities permit an individual to shelter in certain places, or undertake any livelihood activities including those that are livestock-related. Displacement may also have a significant effect on the social and economic networks that people need to reconstruct livelihoods and shelter. Those who can remain in their normal location, despite the impacts of an emergency, may have more security and more access to support.

Interventions will involve decisions on whether displacement/evacuation is necessary or 'remaining in place' is feasible. Where possible, human and livestock shelter and settlement support should be provided to individual households and communities in their original homes. When livestock keepers have been displaced with their livestock, shelter and settlement support should be provided individually or collectively in suitable sites or enclosures. These should be within reasonable distance from grouped settlements for human populations, such as temporary planned or self-settled camps.

Livestock and shelter in different climates

In many areas of the world, temperatures make extreme changes over the course of a year. The need for protection from extreme temperatures may also be accompanied by other adversities: high winds, heavy rain or snow, or extended dry periods. When deciding upon individual shelter, camp and settlement options, it is important to consider to what degree livestock and people traditionally protected or sheltered each other in the local context. For example, they may have shared a cold-weather shelter to share warmth, which may impact the choice and design of the shelter or settlement.

Shelter for livestock can include physical protection from extremes of heat (for example, ventilation or shade), cold (for example, wind barriers), dry weather, or wet weather (for example, raised areas away from or above wet or flooded land). Several common shelter design principles, and shelter materials, work equally well in providing physical protection for both people and livestock. However, enough shelter space or materials must be provided to ensure adequate climate protection for households and livestock at the same time. In the first instance, it is often more cost-effective and easier for the affected communities to use resources already present, such as the shade available under existing trees. This is as long as these resources are equitably available to all who need them.

Option 1: Livestock shelter

Where there is urgent need for livestock shelter, livestock keepers themselves are the first responders and will construct temporary structures. However, where possible, shelter materials and construction should also be adaptable for the long term (see *Table 7.1*). Temporary and longer-lasting livestock shelter interventions may take a range of forms, depending on the needs and nature of the emergency. In almost all instances, these can be combined with related support for human shelter. Actions might include:

- repair, construction, or reconstruction of livestock shelters by contractors or agencies, or directly by affected communities themselves;

Technical standards
Livestock shelter and settlement

- providing materials to livestock keepers for shelter construction – this may include their being provided with support for human shelter construction on the understanding that some materials may also be used for animal shelter;
- incorporating livestock shelter needs into human shelter programming (in other words, designing shelters or camp layouts so that both humans and their livestock can shelter in the same location together);
- providing training in shelter design and construction that covers both combined or separate human and livestock shelter;
- cash or voucher assistance for livestock shelter needs – in many interventions supporting human shelter, agencies use commodity or value vouchers or conditional cash grants (this might involve, for example, their distributing the cash in tranches dependent upon the completion of the shelter in stages, and to approved safety standards: this is to ensure that shelters are safe and meet the needs of everyone in each household); these approaches will need to be aligned with the preferred cash approaches for livestock shelter (see *Chapter 1: Introduction to LEGS* and *Chapter 3: Emergency response planning* on CVA);
- information, advocacy and support for negotiation of secure tenure for land use of individual shelter plots (to avoid forced eviction of livestock and livestock keepers).

There are some types of human shelter, such as 'collective centres', where it is unlikely that the shelter will accommodate livestock. However, for most other types of shelter it is possible for livestock to be accommodated. This will either be in designated structures within the individual shelter plot, or in designated locations separate from shelters but still within a camp or settlement. Where livestock cannot be directly accommodated, negotiation with local host communities is needed to arrange livestock shelter that is near livestock keepers and is safely accessible.

Community livestock shelters can be used for displaced or non-displaced crisis-affected communities. Construction work may include repairing damaged structures or building new structures for groups of households with livestock in spontaneous settlements, collective centres and within camps. Affected communities will often have the skills to build a communal livestock shelter as a group, if given materials and some support. People will often prefer to build multiple smaller community livestock shelters to share with people they know, if they cannot build the preferred individual livestock shelter for each household.

Option 2: Livestock and settlement

Settlement interventions complement livestock shelter construction, particularly for displaced communities. They may include:

- support in negotiations on land rights or on access to grazing and/or shelter or other policy issues;
- liaison with site planners and camp managers about the shelter needs of livestock accompanying displaced populations;
- provision of infrastructure to support the livestock of displaced people (for example, water supply);
- environmental management to address the needs of both livestock and humans in camps in order to ensure public and animal health.

Most settlements that result from emergencies are spontaneous. They are constructed by those who live there, and without the benefit of any initial interventions by trained site planners. As a result, settlements often lack many basic services and have inadequate space for livestock. In such cases, an incremental programme of smaller-scale insertions of basic service facilities is often more realistic than a total reorganisation of the entire settlement. Site planners should be consulted to ensure that expanding a settlement, or inserting new facilities and public spaces, takes realistic account of the needs of livestock and livestock keepers.

Increasingly, both planned and unplanned camps are located in or at the edges of urban areas. These settlements often experience complex challenges in terms of availability of space, interactions with host communities, environmental impact, and access to resources like water, drainage and electrical power. Settlements in or near urban areas may provide livestock keepers with better access to markets, livelihoods, and veterinary support. Yet there may simply not be enough space to provide minimum standards of shelter to all livestock.

Urban and peri-urban shelter and settlement needs fall into three main themes:

- Transhumant herding households affected by emergencies, but retaining their livestock, may conduct their own risk analyses and divide livestock responsibilities. Some household members may become locally displaced and find shelter in urban or peri-urban locations with some livestock (for example, women and girls, so as to access health and other services). Other household members may then keep most of the livestock in rural areas for access to grazing and water.

Technical standards
Livestock shelter and settlement

- Rural households who, having lost all or most of their livestock in an emergency, are forced to migrate to urban or peri-urban areas. This forced displacement will leave families in need of shelter and other resources. Livestock shelter should suit the livestock that people are able to keep in the urban environment. For example, people used to herding sheep in Mongolia who move to a peri-urban area of Ulaanbaatar after a *dzud* may start to raise pigs.
- Urban and peri-urban families affected by rapid-onset emergencies who need replacement for livestock and poultry shelter destroyed during the natural hazard; this might include, for example, replacement of chicken enclosures and pig sheds.

In all these situations, care is needed to ensure that livestock waste and health are factored in to reduce potential disease transmission between livestock and people *(see Chapter 6: Veterinary support)*.

Table 7.1: Benefits and challenges of livestock shelter and settlement options

Option	Benefits	Challenges
1.1 Temporary livestock shelter	This responds to immediate shelter needs of livestock Generally cheaper than longer-lasting solutions, so more people can benefit	It may need to be demolished and rebuilt in the longer term if location, accessibility, or tenure issues are not carefully considered
1.2 Longer-lasting livestock shelter	Livestock keepers remain with a long-term asset after the emergency is over More economical use of resources in the long term	It is generally more expensive than temporary structures Not appropriate for displaced populations who are certain to return to their original areas in a very short period after the emergency

continued over

Technical standards
Livestock shelter and settlement

Option	Benefits	Challenges
2. Livestock settlement interventions	These enable design and planning of wider settlement issues to allow for livestock needs as well as those of their keepers in a range of post-emergency situations. They include both camp and non-displaced contexts They help reduce potential tension or conflict with host communities	Depending on the nature and phase of the emergency, time is limited for discussions with host communities before immediate needs are met Agencies may not recognise the importance of livestock as key livelihood assets for affected communities and may therefore be reluctant to address livestock settlement issues

Timing of interventions

Livestock shelter and settlement needs vary according to the different phases of an emergency. Needs range from immediate response solutions up to durable solutions, once access to these is feasible during recovery. The requirement for livestock shelter may evolve from disaster risk reduction programming, through life-saving interventions, to actions that integrate livelihood recovery with 'building back better'. Shelter and settlement support should prioritise solutions for affected communities that are sustainable in terms of design, location and construction.

It is important to consider the different risks that livestock and livestock keepers face at different times. The threat of robbery or violence will be more prevalent during complex emergencies, as well as in the first chaotic days of a rapid-onset emergency before stability has been established. The risks of forced eviction from a location may become apparent at a later stage. Seasonal cycles are also important. Many communities who become displaced due to an emergency will not be able to re-establish permanent housing for either themselves or their livestock within the first 12 months. In these cases, multiple yearly weather cycles need to be planned for.

Technical standards
Livestock shelter and settlement

Table 7.2: Possible timing of livestock shelter and settlement interventions

Options	Rapid-onset emergency		
	Immediately after	Early recovery	Recovery
1.1 Temporary shelter interventions	✓	—	—
1.2 Longer-lasting shelter interventions	✓	✓	✓
2. Settlement interventions	(✓)	✓	✓

Options	Slow-onset emergency			
	Alert	Alarm	Emergency	Recovery
1.1 Temporary shelter interventions	—	✓	✓	—
1.2 Longer-lasting shelter interventions	✓	✓	✓	✓
2. Settlement interventions	✓	✓	✓	✓

Links to other LEGS chapters and other HSP standards

The provision of livestock shelter complements the livestock interventions described in the other LEGS technical chapters. Where the crisis-affected community is displaced, livestock shelter interventions should be part of a planned response to the full range of livestock needs, including feed, water and veterinary support *(Chapters 4, 5 and 6)*. If livestock are in situations where shelter is vital for their survival and well-being, such as in cold climates, livestock provision interventions *(Chapter 9)* should address shelter needs. In such cases, the intervention should also provide basic advice on animal housing, particularly if species are being introduced to communities that are not familiar with them.

Livestock shelter cannot be considered separately from human shelter and settlement. In some – but not all – instances, both animals and humans will require shelter following an emergency. Coordinated interventions that consider the needs of both humans and their animals will have the greatest impact in the medium and long term, with livelihoods supported and lives saved. However, the settlement needs of communities will always take precedence over those of livestock. It is critical, therefore, that interventions for livestock do not negatively affect the provision of human settlement. But in many cases, settlement needs for humans and livestock are interdependent, further highlighting the need for coordination and joint planning (see LEGS *Principle 5: Coordinated responses*). For example, providing sufficient water to cover the needs of people and livestock is normally considered in a joint needs assessment for a household.

The Sphere Handbook chapter on 'Shelter and Settlements', and the other further reading listed at the end of this chapter, cover human shelter and settlement in detail. The shelter and settlement issues in Sphere that may have a bearing on livestock include shelter design, materials and construction methodologies, land rights, environmental management, and the planning and design of infrastructure such as facilities, buildings and camps.

LEGS Principles and other issues to consider

Table 7.3: Relevance of LEGS Principles to shelter and settlement

LEGS Principle	Examples of how the principles are relevant in shelter and settlements interventions
1. Livelihoods-based programming	Providing shelter and settlement support can reduce livestock mortality and therefore protect the key livelihood assets of livestock owners.
	Use of cash and market approaches to provide livestock shelter can support wider non-livestock owning livelihoods.
	Providing spaces for local markets for livestock products in camps and settlements can help retain livelihoods.
	Shelter locations provide support to additional livelihoods, including their use to store tools related to livestock.

continued over

Technical standards
Livestock shelter and settlement

LEGS Principle	Examples of how the principles are relevant in shelter and settlements interventions
2. Ensuring community participation	It is essential to have community participation in livestock shelter and settlement design. This will ensure human and livestock health and safety, efficiency of responses and cost-effectiveness. Consultation on the location of livestock areas and the construction of livestock shelters is also essential in camps.
3. Responding to climate change and protecting the environment	Livestock shelter and settlement interventions should consider the environmental implications by: establishing climate-appropriate shelter for people and livestock; ensuring sustainable procurement and use of shelter materials; considering the environmental impacts of grazing and water access in and around camps; addressing waste management, including management of livestock-produced waste in camps and settlements; and ensuring environmental rehabilitation in camps after their closure. (This is not low-cost, and in some cases the environmental degradation caused may not be fully reversible or may continue for several years.)
4. Supporting preparedness and early action	Identifying and preparing for the impacts of rapid-onset emergencies (through risk assessments and improving existing shelter and settlement provisions) will mitigate livestock losses. Early action to release tethered livestock is also critical.
5. Ensuring coordinated responses	Coordinated interventions that consider the needs of both humans and their animals will have the greatest impact in the medium and long term, with livelihoods supported and lives saved.
6. Supporting gender-sensitive programming	Shelter and settlement options should support the mitigation of gender-based violence risk, including mitigation of risks for women and children who look after livestock.
	The role of and effectiveness of good site planning can support access to livelihood opportunities for women.

continued over

Technical standards
Livestock shelter and settlement

LEGS Principle	Examples of how the principles are relevant in shelter and settlements interventions
7. Supporting local ownership	Acknowledging the needs, and utilising the expertise, of local host communities in managing livestock-related resources inside and outside a camp is critical.
	Land rights, ethnicity, and local politics may all impact the provision of shelter and settlement. Utilising local knowledge and following conflict avoidance strategies are particularly important when supporting displaced communities.
8. Committing to MEAL	There needs to be more evidence on how specific types of shelter and settlement support for livestock keepers can have a positive impact on livestock keepers and their livelihoods. Monitoring of the interventions, and impact evaluations, can provide valuable information to facilitate learning and improve future practice.

Security and conflict

Rapid-onset natural hazards can threaten the physical security of both humans and livestock. These range from floods and landslides to earthquakes and fires. Complex emergencies on the other hand, cause insecurity by combining civil unrest with slow-onset and rapid-onset emergencies. Livestock shelter interventions may need to consider physical protection from the different types of emergencies and the varied combinations of insecurity. The threat of insecurity may arise or be intensified as a result of the emergency and any subsequent displacement, or it may have been present before the emergency occurred. In all instances, interventions should consider Sphere Protection Principle 1, 'Avoid exposing people to further harm' *(see Chapter 1: Introduction to LEGS).*

To reduce risk, livestock keepers may undertake different coping mechanisms. For instance, self-built walls and barriers around the shelter plots may give better protection for livestock. External interventions for controlling insecurity in camp settings depend on whether livestock keepers prefer to keep their livestock together with their household or physically separated in designated locations used by multiple households at the same time. This in turn depends on local cultures, the amount of space available and the nature of the emergency.

Technical standards
Livestock shelter and settlement

Land use and housing, land and property rights

Lack of secure tenure/rights for accessing housing, land or property can occur when households are displaced into new locations by an emergency. It can also occur when households are not displaced, but when documentation or other proof of tenure agreements has been lost because of the emergency. Alternatively, those who make the decisions about security of tenure (landlords or local authorities) may change as a result of the emergency. This is as much the case for both individual livestock shelters as for livestock shelters in camps or settlements.

In many communities, security of tenure and use of land is dependent on customary agreements rather than on written documentation in full conformity with national laws. In some cases, customary agreement may be stronger and more widely used than 'formal' paper-based mechanisms, and this should be recognised. Displaced communities may not be able to access or engage with either customary or formal land rights mechanisms used by local host communities. Therefore they may not have secure access to land for any livestock, or to land for their own shelter.

When supporting HLP rights for livestock keepers, it may be useful to adapt the 'Due Diligence Standard' approach. This has become commonly used for HLP and for human shelter more generally, and has been adopted by the Global Shelter Cluster. This 'due diligence' approach applies a 'good enough' benchmark, working with local actors to identify security-of-tenure arrangements (formal or informal). These are robust enough to move forward with shelter support, and may be incrementally strengthened later.

Local markets and cash

Working with local markets to provide shelter for livestock can have several benefits. When contemplating a markets-based approach to procuring materials, it is important to coordinate with other humanitarian sectors undertaking emergency market assessments. This should ensure local markets can provide sufficient shelter or construction materials for livestock shelter, as well as for human shelter, for water, sanitation and hygiene (WASH) facilities, and for any other physical structures. Care is needed during the assessment process to ensure that relying on local markets will stimulate the market supply. The markets must not be under too much strain so that they cease to provide other daily items to the emergency-affected and host populations. In addition, while necessary materials may be locally available, they may not be of sufficient quality to provide shelter that is strong enough, durable enough, and gives enough thermal protection. In some cases, restricted commodity vouchers are used that specify the use of construction materials of sufficient quality.

Decision tree for livestock shelter and settlement options

The decision tree *(Figure 7.1)* summarises some of the key questions to consider in determining which may be the most feasible and appropriate option for an emergency shelter and settlement intervention. The standards, key actions and guidance notes that follow provide more information for detailed planning. Where possible, they build on preparedness activities conducted prior to the onset of the emergency/in 'normal' times.

Technical standards
Livestock shelter and settlement

Figure 7.1: Decision tree for shelter and settlement options

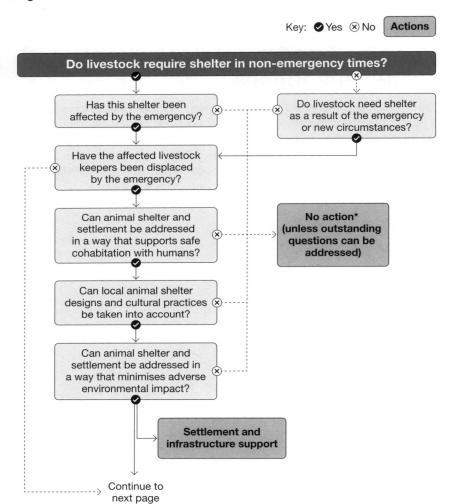

Technical standards
Livestock shelter and settlement

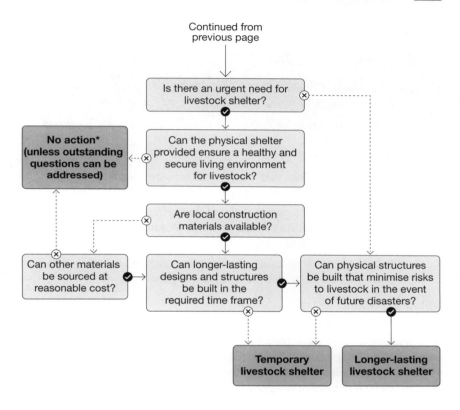

*The result **'No action'** does not necessarily mean that no intervention should take place, but rather that further training or capacity strengthening may be required in order to be able to answer 'yes' to the key questions.

Technical standards
Livestock shelter and settlement

The standards

Standard 1: Preparedness
Livestock shelter and settlement preparedness activities reduce the risks and impacts of emergencies.

Key actions
- Assess the potential risks that rapid-onset emergencies may have on existing livestock shelter provision *(see Guidance note 1)*.
- Promote the design and construction of livestock shelters and settlement planning that will minimise risks to livestock and their keepers and increase resilience to emergencies *(see Guidance note 2)*.
- Identify contingency measures for evacuating confined livestock likely to be impacted by emergencies *(see Guidance notes 3 and 4)*.

Guidance notes
1. Assessment of risks
An assessment of potential risks from the impacts of rapid-onset emergencies and the susceptibility of existing livestock shelter and settlement provision should be undertaken as part of preparedness. Both the number and the density of livestock should be key elements in such assessments.

2. Construction that minimises potential livestock losses
Appropriate construction of livestock shelters and support to settlements can help mitigate the impact of emergencies on livestock. When constructing livestock shelter, mitigation measures to reduce the risk of livestock losses in emergencies may include:

Earthquake resistance: Sites for livestock shelter and settlement infrastructure should always be on stable ground and away from areas at risk of landslides and damage due to aftershocks. Structures for livestock shelter should be designed to be safe in the event of an earthquake through the use of seismic-resistant designs or lightweight construction. While they may use local materials and technology, it may be necessary to advise changes to local building practices to provide for increased earthquake resistance.

Flood impact mitigation: Where possible, livestock shelters should be sited away from areas at risk of flooding, especially flash flooding. Where this is not possible, sites may need improved drainage, or livestock shelters may have to be raised above previous flood levels. Reinforced construction may be considered for foundations to reduce the risk of building failure.

Cyclone resistance: Livestock shelter construction should ensure that roofs are adequately tied and secured to the structure. It should also ensure that structures are located away from the immediate coastline if there is danger of related tidal surges.

Tsunami impact mitigation: Animal shelters should be located away from the coastline wherever possible.

Volcano impact/wildfire mitigation: Lava flows, falling ash and wildfires may be life-threatening to livestock, may damage feed or water supplies, and may also destroy existing livestock shelters. As livestock may be trapped in shelters if they stay there for too long at the start of an eruption or wildfire, early evacuation may be a safer policy. Livestock evacuation plans, developed with at-risk communities, should be prepared in advance.

In all these cases, technical expertise from construction and site planning or urban planning specialists should be sought to ensure the construction adheres to best practices in disaster risk reduction.

3. Contingency measures for evacuation of livestock

Some types of rapid-onset emergencies, such as earthquakes, give no warning and no time for the evacuation of livestock to safety. However, there are other rapid-onset emergencies, such as cyclones, wildfires, floods and volcanic eruptions, where early-warning systems may provide a short time window in which the evacuation of livestock is possible. For these emergencies, communities in at-risk locations can be supported to develop livestock evacuation plans and evacuation enclosures for livestock. The routes should not block the evacuation of the human population, but will need to lead to designated areas close enough to human evacuation centres. This is so that livestock keepers do not risk their lives in order to guard or to feed their livestock at the height of the emergency. Planning for evacuations and enclosures will also need to consider livestock feed and water requirements.

4. Untethering of animals

Another critical approach is identifying and putting measures in place to ensure that livestock are temporarily freed, to avoid the risk of starvation in an emergency before other assistance arrives. Experience has shown that animals such as dairy buffaloes and cows have died where they were

Technical standards
Livestock shelter and settlement

tethered when the families to whom they belonged were killed or injured. A simple response is to untie or release these animals so that they have a chance to find feed and water. These animals should be marked with paint, for example, so that they can subsequently be reunited with their keepers. Emergency preparedness activities should include encouraging livestock keepers to do this in emergencies when sufficient warning is given.

> ### Standard 2: Assessment and planning
> Assessment and planning for livestock shelter and settlement needs are based on community consultation, consideration of local environmental impacts, and sustainable livelihoods programming.

Key actions

- Consult crisis-affected communities on local animal shelter and settlement practices. Ensure that the consultations also include those who have been indirectly affected by the emergency such as host communities *(see Guidance note 1)*.
- Aim to meet the livestock shelter needs of the most at risk in the community *(see Guidance note 2)*.
- Assess the likely local environmental impact of livestock shelter interventions to minimise any adverse impact *(see Guidance note 3)*.
- Ensure the sustainable livelihood needs of the community form part of the assessment and inform the emergency response *(see Guidance note 4)*.
- Where appropriate, conduct a market assessment to investigate the feasibility of cash or voucher transfers to support shelter and settlement interventions *(see Guidance note 5)*.
- Negotiate livestock shelter and settlement interventions with all relevant wider stakeholders *(see Guidance note 6)*.

See *Appendix 7.1: Assessment checklist for livestock shelter and settlement provision*.

Guidance notes

1. Community participation

Livestock keepers need to be consulted on which types of animal shelter are typical for the species they keep, as well as on materials, site selection, site access considerations, hygiene, and livestock management. Special attention should be paid to:

- understanding roles and responsibilities for animal care (age- and gender-based divisions of labour);
- involving host communities when displaced livestock keepers receive support in camp settings (this is to ensure that the location of the livestock shelter and settlement infrastructure does not cause conflict, environmental pressure, or competition for employment or natural resources);
- considering community versus individual shelter options, based on discussions with affected communities, local norms, and the current conditions (security, weather, etc.). In most cases, it is preferable to provide livestock shelter for individual households based on practice prior to the emergency. However, this may not always be possible, appropriate or affordable.

2. At-risk groups

Assessment and planning should examine the specific needs of potentially at-risk groups. They should look at the need for priority assistance to unaccompanied children, the elderly, the sick or those with mobility impairments, who may not be able to build their own livestock shelters. Those without access to construction materials may also need additional assistance. As with any intervention, assistance provided to at-risk groups should not undermine the ability of a community to provide and care for these groups using its own coping strategies.

3. Local environmental impact

Agencies must assess the impact on the local environment of livestock shelters and settlement interventions. Such assessments should include any unsustainable use of local materials or unsustainable concentration of livestock in confined areas. This may be particularly important in camp settings *(see also Standard 4)*. The impact of livestock shelters on the environment can generally be divided into four areas, all of which need to be adequately considered in the initial assessment and planning for livestock shelter, as well as in subsequent monitoring and evaluation:

Technical standards
Livestock shelter and settlement

- the choice of materials, how they are used to construct livestock shelters, and the ways materials no longer needed or exhausted are disposed of;
- the impact of livestock movements, grazing or sheltering on areas in and around camps and settlements;
- the impact of livestock-related waste products (including waste products from slaughter) on areas in and around camps and settlements;
- the impact of zoonotic disease transmission from wildlife or host community livestock and the mixing of animals that previously did not come into contact (for example, in community livestock shelters). Agencies should consider the potential impacts on local human and displaced populations, for example, if bird flu is present in the host poultry population or in the wild bird population.

4. Sustainable livelihoods

Temporary measures to support livestock during an emergency may be required. However, every effort should be made to ensure that shelter and settlement interventions consider the long-term livelihood needs of the affected community. This includes taking into account the impact of anticipated changes to land use, permanent changes to community livelihoods, and changes to livestock management practices as the community recovers from the emergency.

5. Market assessment

Where construction materials are locally available, agencies should assess the possibility of providing cash or vouchers for their purchase. This will support local markets and give greater control over the process to the affected communities. Specialists can help ensure that the required technical specifications for livestock shelter materials are included in market assessments and are followed in livestock shelter construction.

6. Wider stakeholder participation

Livestock shelter interventions should be negotiated with stakeholders beyond the affected community, especially when livestock keepers have been displaced. Stakeholders should coordinate livestock shelter and settlement interventions for displaced populations with human shelter and settlement responses to ensure coherent planning and complementarity of activities *(see LEGS Principle 5: Coordinated responses)*. Stakeholders may include the local authorities that deal with agriculture, water supply, sanitation, land use and housing.

Technical standards
Livestock shelter and settlement

Standard 3: Livestock shelter
Livestock are provided with a healthy, secure living environment appropriate to the context and for the intended use.

Key actions
- Base the design of livestock shelter interventions on local animal housing designs *(see Guidance note 1).*
- Ensure that livestock shelter provides adequate protection from any hazards, prevailing climatic conditions, and the extremes of daily and seasonal weather *(see Guidance notes 1 and 2).*
- Ensure that livestock shelters are constructed using designs and materials that minimise any environmental impact *(see Guidance notes 3 and 4).*
- Design and construct livestock shelter appropriately for the species and use. Even if constructed for temporary use, the materials and structure should be capable of longer-term use or adaptation *(see Guidance notes 2, 3 and 5).*
- Ensure shelters are safely accessible by all livestock keepers *(see Guidance note 6).*
- Establish quarantine systems for community livestock shelters *(see Guidance note 7).*
- Ensure that livestock are afforded adequate physical protection from theft and predators *(see Guidance note 8).*

Guidance notes
1. Appropriate hazard-free living environment
Making livestock shelters safer from earthquakes, fires or floods employs, with few exceptions, the same materials and engineering techniques that are used to make human shelters safer. Agencies should consult with shelter specialists to identify options that use local construction techniques and locally available materials. This will mean that shelters can be sustainably constructed, maintained and repaired by the livestock keepers themselves. It is important for them to incorporate community knowledge on local design and cultural norms, local building materials and local construction methods. Only very rarely will 'shelter systems' or imported prefabricated solutions be appropriate. When deciding on safe locations for livestock shelter, it is important to consult both those with local knowledge of the terrain and those with humanitarian mapping expertise.

2. Healthy, secure living environment

In hot climates, shelter should provide well-ventilated shaded space. In cold climates, it should provide a suitably well-sealed enclosure that is free from draughts and offers insulation from the ground. Where extreme weather conditions prevail, agencies should address shelter needs before the provision of livestock. *(See also Appendix 7.3: Livestock shelter and climate challenges.)*

3. Appropriate design

Shelter for livestock should be based on local building technologies and use local materials. After a natural hazard, livestock shelter may be built using salvage material from damaged infrastructure and buildings. Efforts to maximise the potential for salvage should be encouraged, with toolkits distributed and training provided in their use. See p. 119 of *FAO* (2016), for a table on space requirements and different materials; see also p. 122 for design considerations for livestock shelter.

4. Environmental degradation

If livestock shelter construction requires or encourages the harvesting of locally available material, it can risk permanent environmental degradation. The cutting of trees to provide construction timber for shelters and enclosures, or to fire bricks, is a particular risk. In order to minimise any negative environmental impact, agencies should work with experts and local actors to undertake the following actions:

- Select shelter designs which use materials efficiently, and require the least amount of repair and replacement of materials.
- Select shelter materials that are from identifiable sustainable sources.
- If using 'living' resources, such as thorn bushes, as a main material for livestock enclosure barriers, ensure that the species is native to the local area. Ensure also that it is not invasive, and will not cause any damage to the local ecology.
- Consult with site planners and environmental experts about camp layouts that can disperse both the routes and the enclosures for livestock, to reduce the concentration of environmental impact. This should be where local custom and security concerns permit.
- Consult with all local stakeholders, including from the host community, to identify areas that are the most vulnerable in terms of impact from livestock. These might need barriers or other physical protection installed.

5. Transition to longer-term shelter

Some emergencies may require urgent provision of livestock shelter to ensure the survival of the animals. However, these shelters may not be suitable for the long term. Agencies may need to support communities to reconstruct longer-lasting shelter. The potential to integrate emergency livestock shelters into transitional or more permanent structures is particularly important. For example, designs for livestock shelter for emergency use might incorporate long-lasting roofing and structure in anticipation of a later upgrade to permanent shelter with walls, doors and fencing.

6. Universally accessible design

The design of livestock shelters is only appropriate when livestock keepers can fully use and access it regardless of their gender, age or whether they have any disabilities. Shelters should have dimensions and features that allow entrance and usage by persons with disabilities. In consultation with women and girls, agencies may need to provide additional safety features such as lighting at night-time, to reduce the risk of gender-based violence in and around livestock shelters.

7. Community shelter

In some cases, community shelter may be the preferred option. Before livestock enter the communal shelter, the following activities should be undertaken:

- Identification/marking of livestock should be carried out with the least amount of pain to livestock.
- Livestock should be screened for parasites and disease.
- Livestock health status should be recorded on arrival (for example, their general condition; if they are pregnant).
- Livestock coming from a distance should be quarantined in a separate shelter for a period suitable for local prevalent diseases to become apparent.
- The monitoring of livestock in the quarantine shelter and the communal shelter should be considered. A rota could be set up and roles defined for the households that have livestock in the shelter.
- Depending on local knowledge and structures, quarantining should be integrated with existing environmental and animal health systems and practitioners. This will also help with local acceptance and conflict resolution with host communities, as they can be more confident that livestock arriving in their community are being monitored for disease. (See *Chapter 6: Veterinary support*).

8. Theft and attack

Livestock shelters and settlement interventions should ensure that animals are protected in accordance with local norms from theft and predators. This may include provision of suitable doors with closing mechanisms or secure enclosures around livestock accommodation. Agencies need to consult with livestock keepers, site planners, camp managers and local authorities, about the best security strategies (for example, community security patrols in and around the livestock areas). This should take place whether livestock are sheltered in separate locations or not.

> ### Standard 4: Livestock and settlement
> Settlement supports safe and sustainable cohabitation with humans, and provides a secure, healthy and sustainable environment for livestock.

Key actions

- Ensure that settlement planning and implementation supports human safety and the safe cohabitation of livestock with humans *(see Guidance note 1)*.
- Ensure that settlement planning supports safe access for women and girls, men and boys, to all livestock enclosures, livestock water resources, and other livestock-related infrastructure inside and outside the camp *(see Guidance note 2)*.
- Ensure that livestock-related infrastructure is accessible to livestock keepers with disabilities *(see Guidance note 3)*.
- Ensure that the safety of both humans and livestock has been considered, in terms of site planning for evacuation routes and fire safety *(see Guidance notes 4 and 5)*.
- Minimise the local environmental impact of support to settlement *(see Guidance notes 6 and 7)*.
- Ensure that livestock shelter and settlement activities support sustainable human settlement *(see Guidance note 8)*.
- Ensure that settlement infrastructure minimises negative public health impacts *(see Guidance note 9)*.

Technical standards
Livestock shelter and settlement

Guidance notes

1. Human safety and cohabitation

The location of livestock shelters can affect the safety and protection of livestock keepers. For example, shelters built at some distance from human habitation may expose people to the risk of physical attack, particularly women and children, especially in conflict areas. Conversely, livestock shelter and infrastructure too close to human settlement can increase the risk of spreading disease. This may involve trade-offs concerning the maximum distance between livestock shelters and human settlements for security against theft and attack, and the minimum distance between livestock shelters and human settlements for health and prevention of disease. Agencies should consult with experts on this.

Settlement planning by agencies should also provide for the safe cohabitation of livestock and human communities. This is particularly important to reduce the risk of zoonoses. Site plans should ensure that waste-product treatment areas or drainage networks are separated from those used by humans (including those used by the host community). Greater distancing may be necessary between infrastructure for livestock waste treatment and facilities for human health programming, food storage or education, as well as child-friendly spaces.

2. Mitigation of gender-based violence

Women and girls, and men and boys, involved in livestock-keeping activities have the right to engage in these activities safely and in dignity, and without the threat of gender-based violence (LEGS *Principle 6: Gender-sensitive programming*). In camp settings, agencies should engage with others involved in site planning, and in particular, with women's groups. This process can identify tasks that women and girls undertake for livestock keeping, along with site layouts, public facilities, or infrastructure that could help mitigate livestock-related gender-based violence risks.

3. Accessibility

For members of livestock-keeper households with disabilities, displacement into camps or settlements may pose particular challenges. Agencies should engage with organisations for persons with disabilities, local committees, and shelter and settlement specialists. They can then develop design modifications for livestock-related infrastructure in order to reduce the barriers to livestock-keeping tasks experienced by persons with disabilities.

4. Fire safety and evacuation

A basic requirement of site plans for camps is adequate routes for emergency evacuation and for emergency vehicle access. This is in case of either armed attack on the camp, or the outbreak of fire. As a rule, planners incorporate such routes into an overall site plan, including firebreaks (designated gaps between shelter blocks to prevent fire spreading) and fire extinguisher stations. At a time of sudden emergency inside a camp, livestock keepers may want to make an emergency evacuation of their livestock. Livestock may also attempt to break out of their enclosures in such emergencies. To protect human lives in these situations, it is imperative that livestock should not be able to enter human-designated evacuation routes, emergency access routes, or assembly points. Site planners need to establish alternative routes for livestock, and install barriers or diversions between livestock areas and human evacuation routes. Depending upon the local context, this may be easier if designated livestock enclosures are situated at the edges or rear of the camp, and away from main entrance routes.

5. Pathway encroachment

In interventions, care needs to be taken that, during any dry season, a living fencing enclosure does not become a fire risk. There also needs to be a firebreak gap between the fencing for these enclosures and any human shelter plots. Generally, fencing or enclosure barriers should not encroach upon or block any of the designated pathways for humans in the camp or settlement.

6. Local environmental impact

The environmental impact of support to livestock in settlements should be minimised. Support should also avoid dense concentrations of livestock to reduce the risk of overgrazing. Site plans for camps or settlements may therefore need to include several smaller livestock enclosures, rather than one single central enclosure for all livestock in the camp. In cases where this strategy for multiple dispersed enclosures is necessary, site planners still need to ensure that each enclosure allows access to water and feed resources. It should also be safely accessible by all livestock keepers.

7. Land access negotiations

The inclusion of livestock in camps puts significant additional pressure on the local environment and resources. Because competition with local livestock populations for resources may be a potential source of conflict, interventions must include negotiated access to pasture and grazing with the local population. Negotiated access may change according to the season. It may also need to be rotated between locations, in order to be more sustainable.

8. Sustainable settlement of humans and livestock

Livestock keepers may be remaining in their locations of origin; establishing durable housing or shelter in new locations following displacement; or may be returning to locations of origin after a period of displacement. In all cases, tensions may be present, or increased, over the use of land and other resources for livestock. Where livestock keepers have been displaced, settlement must consider local grazing rights and management structures, accessibility, and land rights and ownership. Resolution is likely to require extensive consultation with stakeholders, as well as advice from local authorities and specialists in other sectors in order to identify sustainable solutions *(see Standard 2, Guidance note 6)*. Agencies may need to support negotiations for conflict mitigation, and safe and sustainable sharing of land resources through appropriate design of housing, settlement and land use zoning.

9. Public health impact

Settlement should be designed to allow for the hygienic management and disposal of animal excreta, especially where livestock-keeping communities are displaced and living in camps. Management options could include:

- providing cash or other incentives for spreading manure;
- ensuring night enclosures do not have a large build-up of dung and are kept clean or moved regularly;
- building enclosures outside the perimeter of human settlements to limit livestock access;
- ensuring adequate distance between human dwellings and animal shelters;
- ensuring that livestock water sources, enclosures and routes are downstream from any water source that is used for human use (this should include any water source intended for schools and child-friendly spaces, or for health facilities);
- ensuring any drainage networks used for livestock-related waste (including waste products from slaughter and livestock markets) are separate or downhill from drainage networks used for human shelter in the same area. This should include any nearby settlements of host communities.

The density of livestock should also remain at a safe level – see Table 14 on p. 119 of *FAO* (2016), for more details on the spatial requirements of different species.

> ## Standard 5: Transition to durable livestock shelter solutions
> Livestock shelter and settlement planning supports a transition from emergency shelter to more durable and sustainable solutions.

Key actions

- Design livestock shelters so that they can be maintained, repaired and upgraded over time by the livestock keepers *(see Guidance notes 1 and 2)*.
- Develop a strategy for settlement planning that considers potential increases in livestock population and livestock variety during the time of displacement into a settlement. It should also take into account the evolving needs of livestock and livestock keepers over that time *(see Guidance note 3)*.

Guidance notes

1. Transitional shelter approaches

Shelters constructed by livestock keepers or humanitarian actors may be very basic and short-term in the first phases after an emergency. Thereafter there may be a need to replace or upgrade them so they can be transitional and play a role in recovery towards durable solutions for affected communities. Depending on the local context, this 'transitional' quality may mean that livestock shelters are designed to be 'transportable' (movable to a different location) or 'transferable' (able to be donated or exchanged among households, depending on needs). They may also be designed as 'transformable' (able either to be upgraded and made more durable or changed so that the shelter is used for a different function by the household). Similar approaches, and design techniques, may already be common for human shelters, depending on the local context.

2. Technical training

As part of the support to transition and upgrade the livestock shelters over time, implementing agencies may consider including technical training for the construction of shelters. This should specifically include likely livestock shelter needs for when livestock keepers are able to return to their locations of origin or establish themselves in new locations.

3. Changing needs for spaces in a camp or settlement

Livestock keepers during a first emergency displacement, or when first entering a camp, may do so with depleted livestock numbers. Over time, the numbers may increase (through natural births, market acquisition or both), and their land-space requirements in and around a camp may also increase accordingly. Further pressure on land use may result from expansions of the camp to provide shelter for an increased human population. Or it may come from competition for land or other resources due to the positive development of other livelihood opportunities (cropping agriculture, see SEADS, or small-industry workshops, etc.) over time. In addition, as livestock shelters themselves are upgraded, the upgrading process may require more space. All of these issues will need to be integrated into the overall plan both for the initial site and for any eventual site expansion. Depending on the local context, an initial strategy of smaller, more dispersed livestock enclosures may help in this transition.

Appendices

Appendix 7.1: Assessment checklist for livestock shelter and settlement provision

Shelter (temporary and longer-lasting)

- Is there an immediate need for temporary livestock shelter?
- What is the estimated population of the different species of animals that may require shelter?
- What specific housing requirements do the different species have in the particular climatic and environmental conditions?
- What are the key social groups?
    a. What are the roles of men and women, girls and boys in particular components of livestock care?
    b. Who in the community is normally responsible for shelter construction?
    c. Are there groups with special needs or who are at higher risk (for example, displaced women or unaccompanied children)?
- What are the prevalent shelter options for the human population? Are they in temporary or transitional shelters, or in older shelters that require repair or reconstruction?
- What are the local animal housing designs, construction techniques, and raw materials?
- Do these building practices adequately reduce the risk of loss in future emergencies?
- Are sufficient local materials available?
    a. How are local construction materials harvested?
    b. Will construction of shelters cause significant environmental destruction?
    c. Would cash or vouchers be appropriate for supporting shelter reconstruction without negatively affecting local markets?
    d. Should building materials be transported into the area?
- What local construction practices can be adapted for the shelter design, in order to make the shelters more resistant to any hazards in the future?
- What practices related to livestock shelter, such as tethering animals rather than giving them free movement, might keepers need to change to reduce immediate livestock mortality in any future event?

Settlement issues

- Are there specific hazards (for example, flooding, landslides, unexploded ordnance – UXO) that are present either in any self-settled locations, or in any potential locations for planned camps? Are there any other 'red line' issues that may make the site non-viable (for example, constant or seasonal lack of sufficient water for both humans and livestock)?
- What are the settlement patterns of livestock keepers?
    a. Have livestock keepers been displaced from their original settlements?
    b. Is there potential for conflict between different livestock-keeping communities (for example, a displaced population and the host community)?
    c. Are there adequate grazing resources locally? Is pasture degradation a potential consequence of the presence of displaced people and their livestock after the emergency?
    d. What are the existing land rights and management systems for communal or shared livestock shelters and settlement infrastructure, and will these be appropriate for any newly constructed shelters?
    e. What other settlement needs do livestock keepers have (for example, safe access for women, boys and girls)?

Technical standards
Livestock shelter and settlement

Appendix 7.2: Examples of monitoring and evaluation indicators for livestock shelter and settlement

	Process indicators (measure things happening)	Impact indicators (measure the result of things happening)
Provision of shelter and settlement support	Number of shelter structures supported by type and location Number and type of settlement interventions	Number of households/livestock with access to shelter versus number of households/livestock in need of shelter Mortality in sheltered livestock versus mortality in livestock without shelter Reported increased or decreased access to livestock products as a result of shelter interventions, particularly for at-risk groups Number of households/livestock with access to grazing, infrastructure and other settlement needs Number of livestock keepers from at-risk groups (women, boys and girls, older people, people with disabilities, households from ethnic or other minority groups, etc.) reporting safe access to livestock shelter

Appendix 7.3: Livestock shelter and climate challenges

Climate challenges	Common livestock shelter interventions
Heat	Roofed shade areas (with higher roofs for additional ventilation)
	Shade walls
	Water tanks/livestock bathing areas
	Camp layout plans that permit cross-ventilation and wind flow around areas used for livestock as well as humans
	Use of shelter materials with slow thermal-transfer properties (e.g. mud bricks)
	Use of shelter materials with high reflective qualities
Cold	Additional insulation for livestock shelters
	Livestock shelter designs with lower roofs, to reduce the volume of space that needs to be heated
	Where culturally appropriate, shared shelter (and shared warmth) between households and livestock
	External walls as wind barriers for outdoor pens and paddocks
	Sufficient indoor storage space for livestock cold-weather feed
Dry/reduced water supply	Additional space inside shelters, or in shaded areas, for water-storage containers
	Construction of watering points, with wells or boreholes or water reservoirs designated for livestock, in locations safely accessible by all in and around camps or settlements (see *Chapter 5: Water*).

continued over

Technical standards
Livestock shelter and settlement

Climate challenges	Common livestock shelter interventions
Wet	Raised foundations or plinths for livestock shelters, to provide dry standing space and feed-storage areas
	Livestock shelters designed with roofs with wide overhangs, to provide additional protection from rain both in and around the shelter
	Integration into the main drainage networks in camps of drainage and waste management systems for livestock-use areas.

(Note: The design of the interventions listed in this table should be undertaken by agencies in consultation with experts and affected communities.)

References and further reading

Camp Coordination and Camp Management (CCCM) Cluster (2021) *Minimum Standards for Camp Management*, CCCM Cluster, https://cccmcluster.org/resources/minimum-standards-camp-management

Catholic Relief Services (2012) *Managing Post-Disaster (Re)-Construction Projects*, How-To Guide, CRS, Baltimore, MD, https://www.crs.org/our-work-overseas/research-publications/managing-post-disaster-reconstruction-projects

Corsellis T. and Vitale A. (2005) *Transitional Settlements: Displaced Populations,* Oxfam GB and University of Cambridge shelter project, https://www.sheltercluster.org/sites/default/files/docs/Transitional%20Settlement%20-%20Displaced%20Populations.pdf

Dosti Development Foundation (DDF) and FAO (2007) *Livestock shelter and supplementary cattle feed project report, 2006–2007*, DDF and FAO, Pakistan.

FAO (2016) *Livestock-related interventions during emergencies – The how-to-do-it manual.* Edited by Philippe Ankers, Suzan Bishop, Simon Mack and Klaas Dietze. FAO Animal Production and Health Manual No. 18. Rome, https://www.fao.org/3/i5904e/i5904e.pdf (See Chapter 8)

International Federation of Red Cross and Red Crescent Societies (IFRC) and Oxfam (2007) *A Guide to the Specifications and Use of Plastic Sheeting in Humanitarian Relief*, IFRC and Oxfam, https://www.sheltercluster.org/sustainable-solutions-working-group/documents/plastic-sheeting-guide

International Organization for Migration (IOM) and The Shelter Cluster (2018) *Site Planning: Guidance to Reduce the Risk of Gender-Based Violence*, 3rd edn, Global Shelter Cluster, https://sheltercluster.s3.eu-central-1.amazonaws.com/public/docs/site_planning-gbv_booklet_apr-2018_web_high-res_v3_0.pdf

IOM, Norwegian Refugee Council (NRC) and United Nations High Commissioner for Refugees (UNHCR) (2015) *Camp Management Toolkit*, IOM, NRC, UNHCR, https://www.refworld.org/pdfid/526f6cde4.pdf

Jha, A.K. with Barenstein, J.D., Phelps, P.M., Pittet, D., Sena, S. (2010) *Safer Homes, Stronger Communities: A Handbook for Reconstructing after Natural Disasters,* World Bank, Washington, D.C. http://hdl.handle.net/10986/2409

The Shelter Centre (2012) *Transitional Shelter Guidelines*, Shelter Centre, Geneva, https://sheltercluster.org/resources/documents/transitional-shelter-guidelines

The Shelter Cluster (2013) *Land Rights and Shelter: The Due Diligence Standard*, Shelter Cluster, https://sheltercluster.s3.eu-central-1.amazonaws.com/public/docs/4.2_gsc_land_rights_and_shelter_the_due_diligence_standard.pdf

Sphere Association (2018) *The Sphere Handbook: Humanitarian Charter and Minimum Standards in Humanitarian Response*, 4th edn, Geneva, https://spherestandards.org/handbook-2018/

UNHCR (2015) *UNHCR Emergency Handbook*, 4th edn, UNHCR, https://emergency.unhcr.org/topic/27459/camps

UNHCR and International Union for Conservation of Nature (IUCN) (2005) *Planning and Animal Husbandry in Refugee and Returnee Situations: A Practical Manual of Improved Management*, UNHCR and IUCN, Geneva, http://www.unhcr.org/protect/PROTECTION/4385e3432.pdf

In addition, many practical case studies for shelter and settlement can be found in the various editions of Shelter Projects, https://www.shelterprojects.org/

See also case studies for livestock shelter and settlement interventions in emergencies at: https://www.livestock-emergency.net/resources/case-studies/

Technical standards
Livestock offftake

Chapter 8: Technical standards for livestock offtake

- 290 Introduction
- 292 Options for livestock offtake
- 296 Timing of interventions
- 297 Links to other LEGS chapters and other HSP standards
- 298 LEGS Principles and other issues to consider
- 301 Decision tree for livestock offtake options
- 303 The standards

- 315 Appendix 8.1: Assessment checklist for livestock offtake
- 317 Appendix 8.2: Examples of monitoring and evaluation indicators for livestock offtake
- 319 References and further reading

Photo © FAO/Mohammad Rakibul Hasan

Technical standards
Livestock offtake

Chapter 8: Technical standards for livestock offtake

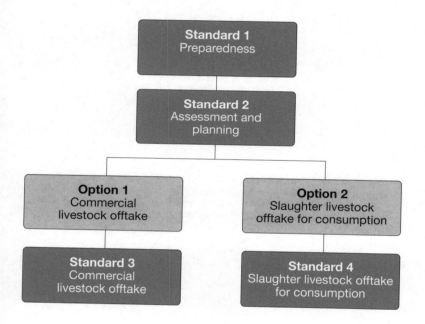

Introduction

Livestock offtake can be critical for enabling livestock-dependent communities to reduce the adverse effects of drought, and to rescue some value from livestock assets in the form of much-needed cash or meat. Livestock offtake is best applied during slow-onset emergencies such as drought, before animals lose their value, die or pose a risk to public health. It is generally less applicable in rapid-onset emergencies.

If livestock offtake interventions are to be effective, preparedness is essential. Three scenarios highlight the importance of advance preparedness when planning livestock offtake interventions:

1. Early interventions in slow-onset emergencies, before livestock become too emaciated, will allow communities to sell their stock at near market values to external buyers.
2. When interventions happen late in a drought, due to lack of preparedness, the prices obtained for emaciated livestock will be very low. Here, livestock are likely to be used for local meat distribution only.
3. Interventions that take place very late are likely to lead to the slaughter of stock for their disposal only *(see Box 8.1)*.

This chapter presents information on livestock offtake as a livestock emergency response. It also provides the technical options for livestock offtake interventions and the associated benefits and challenges of each. Information is also available in *FAO* (2016, Chapter 4: Destocking). For each technical option, LEGS provides information through standards, key actions and guidance notes. Checklists for assessment, as well as monitoring and evaluation indicators, are presented as appendices at the end of this chapter. Further reading is also provided. Case studies are presented on the LEGS website (see https://www.livestock-emergency.net/resources/case-studies/).

Links to the LEGS livelihoods objectives

Livestock offtake can provide immediate assistance to affected communities, protecting their remaining livestock in line with the first and second of the LEGS livelihoods objectives. Livestock offtake can support crisis-affected communities to:

- **obtain immediate benefits using existing livestock assets** – by providing cash or food from the sale of at-risk or unmarketable animals;
- **protect key livestock assets** – by ensuring the survival of remaining animals. Cash sales enable herders to buy livestock feed and water (or prophylactic drugs) or transport remaining animals to less affected areas.

The importance of livestock offtake in emergency response

Livestock offtake is a valuable response to drought when animals would otherwise deteriorate and die. It allows potential livestock losses to be converted into cash or meat. Prior advocacy work may be required ahead of an emergency, when herders may be reluctant to part with the stock that are likely to perish. Livestock offtake can also be viewed as a resilience-building

**Technical standards
Livestock offtake**

tool: removing animals relieves pressure on scarce feed, grazing and water supplies, to the benefit of the remaining stock (see LEGS *Principle 3: Climate change and the environment*). In addition, meat from slaughtered animals supplements the diets of at-risk families, providing much-needed protein in lean periods.

Livestock offtake also contributes to at least two of the animal welfare 'five domains', as described in *Chapter 1: Introduction to LEGS*, which are their 'health' and 'mental state'. When sold animals are moved to a more favourable location, this may allow them to resume their normal behaviour. When necessary, slaughter offtake will also relieve animals from the pain and distress associated with starvation and thirst. As livestock offtake involves handling, transporting and slaughtering animals, it is important to ensure that they do not suffer pain, fear or distress.

Livestock offtake is not usually applicable in rapid-onset emergencies, such as earthquakes and floods, since livestock are either killed or they survive. However, when natural hazards, such as cyclones or fires, destroy available feed supplies, removing animals may be an appropriate response. When animals pose a risk to public health, they need to be humanely slaughtered and disposed of – see *Box 8.1*.

Options for livestock offtake

The two most common livestock offtake interventions are commercial offtake and slaughter offtake for consumption.

Option 1: Commercial livestock offtake

In commercial livestock offtake, agencies provide external support so that market operators (livestock traders, feedlot owners, butchers, etc.) can still continue to operate when the livestock market begins to fail. Market failure can result from the following circumstances: weak demand; excess supply; animals in poor condition; the inaccessibility of animals; or the unwillingness of livestock keepers to sell. The result is usually a collapse in livestock prices and market operators withdrawing from the market. The aim of commercial livestock offtake is to help the livestock market to function in these difficult circumstances. It assists with the marketing of drought-weakened livestock before their condition and value deteriorate too far and they become impossible to sell.

Technical standards
Livestock offtake

There are several benefits to commercial livestock offtake:
- It provides cash for the crisis-affected communities.
- It promotes a longer-term relationship between buyers and livestock keepers.
- It can have an overall positive impact on large numbers of livestock and their keepers.
- It is one of the more cost-effective drought interventions, since it does not involve agencies purchasing animals directly.

To succeed, commercial livestock offtake requires an active private trading system for livestock, and an accessible domestic or export demand. Animals do not always go directly to an abattoir but may be sent elsewhere to regain their condition. They may then be slaughtered or resold later.

Typical agency support to livestock traders can include assistance in bringing together buyers and sellers of animals; and facilitating short-term credit, subsidies and tax exemptions. Bringing livestock keepers and traders together is the simplest and most effective intervention. In order not to disrupt the normal market, the support provided should be the minimum required to facilitate and overcome the immediate constraints.

Some agencies have intervened directly to purchase animals in emergencies rather than working with the commercial livestock traders. Caution is required with this approach, however, to make sure that direct purchasing does not undermine the long-term sustainability of the private market. If traders are still actively operating in livestock markets, interventions are not recommended.

Option 2: Slaughter livestock offtake for consumption

Unlike commercial livestock offtake, agencies rather than private traders initiate slaughter livestock offtake for consumption. It is appropriate when the local market for livestock has failed and traders have withdrawn. Invariably the animals are in a poor condition and livestock prices have collapsed. In these circumstances, an agency purchases animals and arranges for their humane slaughter. Fresh meat is then distributed to the crisis-affected communities. Because fresh meat is perishable, if it cannot be distributed straight away, immediate action must be taken to preserve it by salting, boiling or drying.

Technical standards
Livestock offtake

Slaughter livestock offtake for consumption is a more costly option for agencies than commercial livestock offtake as it involves the direct purchase of animals. The cost is partly offset by the additional benefits from the meat distribution, as well as the employment opportunities – including from the processing of hides and skins. Operational costs can be reduced if agencies directly distribute the purchased animals to target communities for them to undertake the slaughter and meat distribution process themselves. There are also animal welfare and public health benefits associated with improved slaughter and meat processing. In some countries, religious and cultural issues will need to be considered for the slaughter of cattle.

Participants in slaughter livestock offtake for consumption might include:

- those eligible to sell animals for slaughter, especially female-headed households and those from marginalised communities;
- those eligible to receive meat, especially large families, single-parent and orphan households, the elderly and other at-risk groups – if there is enough, it may be simpler to distribute the meat equally to the whole community to avoid potential resentment; agencies often give the meat to another agency for distribution as part of a broader food assistance programme, which may include schools, hospitals and prisons;
- those who may be employed in the slaughter and processing of animals. This will provide them with income as well as skills for the future.

Box 8.1
Slaughter for disposal

In severe and prolonged drought situations, animals may become so emaciated or diseased that they are unfit for human consumption. A decision to conduct slaughter for disposal might then be made by the relevant veterinary or public health authorities based on ante- and post-mortem inspections. In such cases, slaughter should meet locally acceptable welfare standards, and the carcasses must be disposed of to minimise risk to public health.

In some rapid-onset emergencies, such as typhoons, floods and earthquakes, euthanasia of injured or unrecoverable animals (companion as well as other animals) may have to be considered and suitable methods identified.

Considerations for carcass disposal are discussed in *Chapter 6: Veterinary Support*, and in the FAO (2020) carcass management guidelines. FAO (2001) also provides guidelines for humane handling, transport and slaughter of livestock.

The benefits, challenges and key requirements of the different options are summarised in *Table 8.1*.

Table 8.1: Benefits, challenges and key requirements of livestock offtake options

Benefits	Challenges	Key requirements
1. Commercial livestock offtake		
Provides cash for immediate needs and/or reinvestment in livestock	Has to be carried out before stock deteriorate significantly	Interested buyers
		Willing sellers
Builds on existing coping strategies	Traders' preference is for profitable animals; they may not necessarily target emaciated animals owned by at-risk groups	Accessible domestic or export markets
Relieves pressure on scarce feed/grazing and water supplies		Infrastructure: roads; holding grounds; feed and water; trekking to the nearest road; and security
	Carries potential risks to animal welfare through inappropriate handling and transport.	
Can handle large numbers of animals		Conducive attitude within agencies to livestock trade and credit provision
Is relatively low in cost (majority of costs borne by traders)		
		Willingness within agencies to engage with the private sector
Promotes longer-term relationships between buyers and sellers		

continued over

Technical standards
Livestock offtake

Benefits	Challenges	Key requirements
2. Slaughter livestock offtake for consumption		
Provides cash for immediate needs and/or reinvestment	Operationally more complex unless target communities directly engage in the slaughter and distribution of meat	Local institutions to organise, manage and help target communities
Relieves pressure on scarce feed/grazing and water supplies	Higher administration costs in the absence of community participation	Coordination between implementing agencies to agree on methodologies and, in particular, pricing strategies and operational areas
Provides supplementary food assistance	More expensive, as it includes the purchase of animals	Food assistance operations willing to accept meat
Surplus fresh meat can be preserved	Less long-term sustainability	Implementing agency with organisational capacity to manage the programme
Provides employment opportunities within local community	Less conducive to handling larger number of animals	Slaughter infrastructure is already available, or there is potential for construction
	Basic necessities, such as water, may be scarce during drought	Conducive public health policy
		Slaughter and distribution managed by agency in a way that is appropriate to cultural norms
		Ability to use humane handling and slaughter processes

Timing of interventions

The stage of the emergency usually determines the type of livestock offtake undertaken. It is important to note that there is a limited time period for commercial livestock offtake. It is most effective in the alert and alarm phases in a slow-onset emergency (see *Chapter 1: Introduction to LEGS* on emergency phases). Useful preparedness strategies might involve the prior listing of potential livestock traders, value adders, meat processors, ranchers, etc. to allow timely support to commercial livestock offtake *(see LEGS Principle 4: Preparedness)*. Slaughter livestock offtake for consumption, however, invariably takes place in the late alarm, emergency, or early recovery phases. By this time, livestock are in such poor condition that they are unmarketable, and commercial livestock offtake is not applicable. Livestock offtake can also take place after rapid-onset emergencies – for example, in places affected by wildfires, if there is loss of pasture *(see Table 8.2)*.

Livestock keepers rarely value their animals solely in financial terms. They consider many factors, including the chance of their animals surviving in the prevailing conditions. At the height of a drought, they may be willing to sell animals at almost any price, but at the first signs of rain they may change their minds. Agency flexibility is needed to respond quickly to changing circumstances and to switch resources into alternative interventions. The Minimum Standard for Market Analysis (MISMA) may be useful for ensuring prices being offered for livestock remain valid in emergency contexts where inflationary pressures are acute.

Table 8.2: Possible timing of livestock offtake interventions

Options	Rapid-onset emergency		
	Immediately after	Early recovery	Recovery
1. Commercial livestock offtake	✓	—	—
2. Slaughter livestock offtake for consumption	✓	—	—

Options	Slow-onset emergency			
	Alert	Alarm	Emergency	Recovery
1. Commercial livestock offtake	✓	✓	—	—
2. Slaughter livestock offtake for consumption	—	✓	✓	—

Technical standards
Livestock offtake

Links to other LEGS chapters and other HSP standards

An important aim of a livestock offtake intervention is to improve the survival chances of the remaining livestock, especially the core breeding animals. Agencies therefore often undertake livestock offtake with other LEGS interventions as part of an integrated approach. Typically, these include providing feed, water and veterinary support *(see Chapters 4, 5 and 6)*. The LEGS Participatory Response Identification Matrix (PRIM), described in *Chapter 3: Emergency response planning*, is a valuable tool in making these assessments. *Chapter 6: Veterinary support*, contains information on the disposal of carcasses.

Commercial livestock offtake requires a good understanding of market processes. The Emergency Market Mapping Analysis (EMMA) tool and the Minimum Standard for Market Analysis (MISMA) provide standards and guidance to support this process *(see Sources of specific information* in *Chapter 1: Introduction to LEGS)*.

After a drought, rebuilding stock numbers to levels that can sustain a household can take years. In pastoralist and agro-pastoralist communities, livestock interventions alone may not be enough. Additional humanitarian support, such as food assistance, may be required. The Sphere Handbook provides detailed guidance on this.

LEGS Principles and other issues to consider

Table 8.3: The relevance of the LEGS Principles to livestock offtake

LEGS Principle	Examples of how the principles are relevant in livestock offtake interventions
1. Supporting livelihoods-based programming	Livestock offtake supports livelihoods by improving the survival chances of the remaining livestock, especially core breeding animals.
	When livestock offtake is combined with other LEGS interventions (feed, water provision and veterinary support) as part of an integrated approach, livestock livelihoods are further supported.

continued over

Technical standards
Livestock offtake

LEGS Principle	Examples of how the principles are relevant in livestock offtake interventions
2. Ensuring community participation	Ensuring community participation is essential for the fair selection of livestock offtake participants. This selection should be based on agreed criteria and recent risk and capacity assessments.
	Private traders aim to maximise profit and may exclude communities with poor access, poor security or inadequate facilities. Agency assistance given to livestock traders should therefore be conditional on ensuring active community participation, so that those who are at risk are not excluded. However, traders should have the discretion to buy the animals they prefer.
	Direct engagement of communities in the slaughter and distribution of meat helps to reduce the cost of livestock offtake operations.
3. Responding to climate change and protecting the environment	Livestock offtake issues can have positive or negative environmental implications. Slaughtering of animals generates local waste (including condemned carcasses) that needs to be disposed of safely to avoid pollution. Tanning of hides and skins has similar issues.
	The removal of large numbers of livestock can relieve localised pressure on natural resources during a time of scarcity, such as a drought.
	Concentration of animals around camps and markets may have a short-term detrimental effect on the immediate environment.
	Where indigenous breeds are under threat, care should be taken not to exacerbate any loss of local biodiversity.

continued over

Technical standards
Livestock offtake

LEGS Principle	Examples of how the principles are relevant in livestock offtake interventions
4. Supporting preparedness and early action	The time frame for commercial livestock offtake is very limited during a crisis, and preparedness is essential. Prior listing of potential livestock traders, butchers, value adders, etc. when a crisis is imminent, will help to kick-start commercial livestock offtake in the alert/alarm phase.
	In drought, early interventions before livestock become too emaciated allow communities to sell stock at close to their market price to external buyers. Later interventions result in emaciated livestock that will fetch only low prices for meat distribution. Interventions that take place very late result in the slaughter of stock for disposal at minimal prices.
5. Ensuring coordinated responses	The establishment of coordination groups that involve key stakeholders, and affected community representatives, can help ensure that livestock offtake interventions are appropriately targeted and managed.
	Coordination among those engaged in slaughter livestock offtake for consumption is vital. This ensures that purchase prices and ways of working are agreed, and avoids competition and confusion.
6. Supporting gender-sensitive programming	Gender-sensitive programming is necessary when choosing livestock offtake options and for selecting those who should benefit most.
	Slaughtering may provide employment options for men and women (depending on cultural norms). Meat distribution options and drying activities may also provide employment opportunities.
	Cash from selling big animals (cattle, camels, yaks) may increase male spending power. Meanwhile, women may only have the discretion to sell small stock (sheep and goats), as well as poultry, and to control the proceeds from this. Interventions should therefore assess balancing the types of animals selected for livestock offtake in order to benefit both women and men.
	Particular attention is needed to ensure that widows and female-headed households are not excluded from the intervention.

continued over

Technical standards
Livestock offtake

LEGS Principle	Examples of how the principles are relevant in livestock offtake interventions
7. Supporting local ownership	Livestock-owning communities frequently have their own local coping strategies for responding to emergencies. Building on these local strategies through community management of the livestock offtake process (for example, local coordination groups overseeing targeting, pricing and distribution) will increase their effectiveness. Customary offtake practices also observe and respect cultural norms and taboos related to selection criteria, slaughter and meat distribution practices.
	Livestock offtake stimulates the local economy during a crisis through the sale of animals, enabling communities to purchase food and other essential commodities. Livestock offtake operations also provide opportunities for employment of community members.
	When dealing with external traders, local communities and local market actors may need additional support from external agencies.
8. Committing to MEAL	Understanding livestock markets and trends is key to planning and implementing successful livestock offtake. The purchase of livestock for either commercial offtake or slaughter offtake for consumption is likely to have an impact on local markets. Monitoring livestock prices is important to understand this impact and ensure that at-risk groups are not adversely affected.

Decision tree for livestock offtake options

The decision tree *(Figure 8.1)* summarises some of the key questions to consider in determining the most feasible and appropriate option for an emergency livestock offtake intervention. The standards, key actions and guidance notes that follow provide more information for detailed planning. Where possible, these build on preparedness activities conducted prior to the onset of the emergency/in 'normal' times.

Technical standards
Livestock offtake

Figure 8.1: Decision tree for livestock offtake options

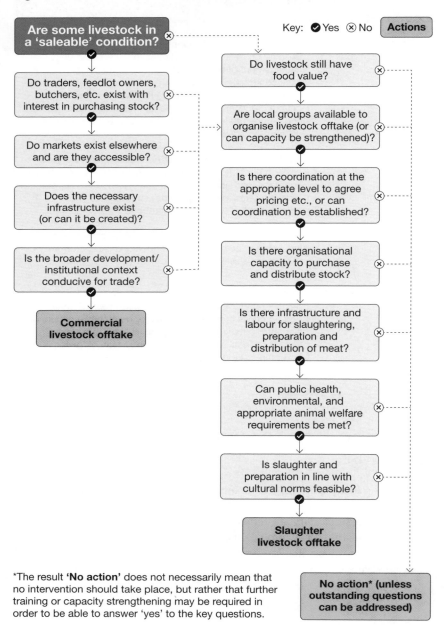

*The result **'No action'** does not necessarily mean that no intervention should take place, but rather that further training or capacity strengthening may be required in order to be able to answer 'yes' to the key questions.

Technical standards
Livestock offtake

The standards

Livestock offtake enables livestock keepers to salvage some value from animals that, without intervention, may have little or no value. Commercial livestock offtake is only feasible before animals lose condition and market prices collapse. Beyond this point, slaughter livestock offtake for consumption may be the only alternative. Preparedness and early analysis of the situation, with the participation of affected communities, is essential for deciding whether livestock offtake is a feasible and appropriate response. A livestock offtake assessment checklist is presented in *Appendix 8.1*.

> **Standard 1: Preparedness**
> The critical time frame for livestock offtake is identified and preparedness activities undertaken.

Key actions

- Monitor relevant early warning data *(see Guidance note 1)*.
- Understand market processes, and monitor the livestock market closely *(see Guidance note 2)*.
- Monitor livestock condition and welfare closely *(see Guidance note 3)*.
- Implement livestock offtake preparedness activities with key stakeholders *(see Guidance note 4)*.
- Encourage the early marketing of livestock *(see Guidance note 5)*.

Guidance notes

1. Early warning

Most drought-affected areas have some form of early warning indicators that can alert agencies that they should start to consider livestock offtake.

2. Monitoring livestock markets

Increased numbers of animals for sale without a corresponding increase in demand, or falling livestock prices, may indicate the beginning of 'distress sales'. This is where livestock keepers are trying to salvage some value from their animals through the normal market.

Technical standards
Livestock offtake

3. Monitoring livestock condition

Deteriorating livestock condition may be an indicator of impending crisis. Participatory planning with communities and local knowledge, coupled with secondary data from early warning monitoring systems, can help determine if the condition of animals is worse than usual for the time of year.

4. Preparedness activities

Preparedness activities might involve prior listing of potential livestock traders, value adders, meat processors, ranchers, etc. to initiate commercial livestock offtake on time. Prior advocacy work on commercial livestock offtake may be required ahead of an emergency. This is because herders may be reluctant to part with the stock that is likely to perish.

5. Timing and early action

The options for livestock offtake relate to the phase of the emergency (see Table 8.2). Encouraging communities to sell stock when prices are still favourable will ensure the greatest livelihood benefits. Pre-positioning of funding sources as 'crisis modifiers' will be needed for early and effective engagement with market processes.

Standard 2: Assessment and planning

The type of livestock offtake activity selected is appropriate to market conditions and the state of the livestock.

Key actions

- Undertake commercial livestock offtake only when traders are willing to buy, and animals are in a suitable condition (see *Guidance note 1*).

- Ensure livestock offtake involves appropriate species, age and type of animal, depending on local circumstances, knowledge and practices (see *Guidance note 2*).

- Assess if customary livestock offtake is still practised in some communities in order to target less fortunate households (see *Guidance note 3*).

- Ensure assessments consider the broader development and institutional context of the emergency (see *Guidance note 4*).

- Ensure the affected communities are fully involved in planning and assessing activities (see *Guidance note 5*).

Technical standards
Livestock offtake

- Assess the security situation to ensure the safety of livestock, their keepers and agency staff *(see Guidance note 6)*.
- Prepare exit strategies in advance *(see Guidance note 7)*.

Guidance notes

1. Commercial livestock offtake or slaughter livestock offtake for consumption

A 25 per cent drop in livestock prices or a 25 per cent increase in the cereal–livestock price ratio is commonly regarded as a trigger point for initiating livestock offtake. Implementing agencies should assess livestock condition and market interest to determine whether commercial livestock offtake is viable.

2. Selection of animals

Removing cattle, camels and yaks has the greatest impact on the immediate environment and injects the most cash into the local economy. However, with such big animals, there are equity and gender issues to consider, since at-risk groups or women may be excluded from the benefits. The inclusion of sheep and goats may allow at-risk groups and women to benefit as well. As a general principle, young female breeding stock should be excluded as they are required for rebuilding the herds/flocks of the future. See *FAO* (2016, Chapter 4), on 'Which animals to include?' as well as LEGS *Principles 1: Livelihoods programming* and *6: Gender-sensitive programming*.

3. Customary livestock offtake practices

In some communities of Eastern Africa (such as the Gabra, Boran and Gari), customary chiefs give an order for the slaughter of a bull or a camel by each capable household in the village in times of emergency. The meat is shared with poor households, and is dried and preserved in ghee to be used sparingly for the lean period. The bones are also kept and cooked repeatedly to provide some fat (marrow). Although this practice may be in decline for various reasons, households that can afford to do so in each community are still likely to practise this coping strategy. Assessing and identifying those villages where such customary livestock offtake practices exist in the community should be undertaken, to promote community-initiated and community-managed coping strategies. This will lessen the involvement of external agencies where such customs are practised.

4. Development and institutional context

The broader context of the emergency needs to be understood to ensure that the risks and opportunities associated with livestock offtake are identified. Relevant information may include:

- any restrictions on cross-border trade and the internal movement of livestock; licensing/tax regimes; existing meat control regulations as stipulated in the laws of each country; access to credit and money transfers; public health and veterinary regulations; and infrastructure;
- assistance provided by other agencies to ensure activities are coordinated and do not overlap with each other;
- policies of the implementing agency, which may regulate its involvement with the private sector or with credit provision, as well as how it can acquire animals or local services.

5. Community involvement

Arrangements for community involvement should be established. This is usually a coordination group of key partners, people affected (including women, young people, the elderly and marginalised groups), representatives of the local authorities, and other agencies operating similar schemes.

6. Security

The extent to which livestock offtake may aggravate existing security problems needs to be assessed. Agencies have a responsibility to protect and ensure the safety of their staff and contractors. They should explore alternatives to carrying cash – such as vouchers, mobile money payments or pre-paid cards.

7. Exit strategies

To ensure livestock offtake has no long-term adverse consequences, it is important to plan how and when operations will finish. Flexibility is needed to accommodate sudden changes in circumstances (market prices, condition of animals, onset of rain, etc.). These could affect the willingness of livestock keepers to sell animals, or traders to participate in the market.

Standard 3: Commercial livestock offtake
Support is provided for selling marketable animals.

Key actions
- Involve the affected communities *(see Guidance note 1)*.
- Assess demand for meat and animals, and identify weaknesses in the supply chain *(see Guidance note 2)*.
- Identify key partnerships *(see Guidance note 3)*.
- Select areas for intervention, taking account of available animals, infrastructure, and security *(see Guidance note 4)*.
- Assess the specific categories of livestock buyers and their preferences *(see Guidance note 5)*.
- Agree and publicise criteria for selecting animals and setting pricing guidelines (see *Guidance note 6*).
- Assess transaction costs *(see Guidance note 7)*.
- Identify support that is essential for the success of the intervention *(see Guidance notes 8 and 9)*.
- Provide and monitor essential ongoing support *(see Guidance note 10)*.

Guidance notes

1. Consultations and coordination groups
The aim of a coordination group is to oversee and evaluate activities and to ensure that at-risk people are not excluded. The group should also act to pre-empt and resolve disputes. Participation of trader representatives is essential *(see also Standard 2, Guidance note 5)*.

2. Livestock market and supply chains assessed
There must be a demand to absorb the extra animals entering the market as a result of a livestock offtake initiative. This may be a terminal (domestic or export) market, or an intermediate market for holding or fattening weakened animals. Information may be available from government or parastatal departments, such as the ministries of agriculture and trade, the statistics office, etc. This can include information on prices, the number of animals sold, supply and demand patterns, market facilities, and trade networks.

3. Partnerships

Successful commercial livestock offtake depends on partnerships between the implementing agency and private livestock traders. Trade associations may assist in identifying and listing suitable partners. Where possible, a core group of committed partners should be identified who have the interest and capacity to lead the initiative *(see Impact case study: Commercial livestock offtake in Ethiopia)*.

4. Intervention areas

Selection of appropriate locations for commercial livestock offtake should be based on assessments of:

- the prevailing security situation as it affects traders, livestock keepers and agency staff;
- a sufficient supply of animals for sale;
- livestock traders willing to buy;
- suitable infrastructure: roads, temporary markets, holding grounds, etc.;
- veterinary restrictions on moving animals.

5. Categories of livestock buyers

In general, commercial livestock offtake operations attract livestock buyers with different interests. Domestic live animal traders, butchers and meat processors tend to prefer to buy animals in reasonable condition for selling in other profitable markets or for the meat industry. This group of buyers is likely to offer better prices, as the animals are not yet too weakened. Exporters and feedlot operators (and ranchers, if present) tend to prefer drought-weakened but large-framed animals. This is so they can recondition them in a short period and make substantial profits. This group prefers to conduct transactions during the emergency phase of a drought. Agencies should take note of this dynamic when partnering with traders.

6. Livestock selection and pricing

Commercial livestock offtake aims to help the normal market to function in difficult circumstances. Ideally, it also establishes new and continuing relationships between livestock keepers and traders. The species and types of animals purchased should be more or less like those marketed under normal conditions – generally surplus males. The prices paid for livestock supported by commercial offtake should be agreed within the coordination group (see *Guidance note 1* above) to ensure transparency and fairness.

7. Transaction costs

Fees for markets, movement permits, abattoirs and meat inspection are transaction costs usually borne by the trader. If these costs are too high, they may restrict trade from happening in the more remote markets, or trade of animals in poorer condition. These fees are also important sources of income for often cash-strapped local institutions. The use of external support to pay these fees directly to local institutions may be preferred to the temporary suspension of trading.

8. Appropriate support

It is important for agencies to understand the critical constraints and weaknesses when markets are under stress, so they can identify appropriate support. In order not to disrupt the normal market, agencies should provide the minimum support required to facilitate and overcome the immediate constraints. Support may include agencies:

- bringing interested traders and livestock keepers together. This can be done by them organising and publicising temporary markets and by providing holding facilities, additional security arrangements, on-site feed and water, arbitration services, etc.;
- providing credit (or facilitating access to credit) for traders to purchase animals;
- supporting transport costs to remote areas. Fuel subsidies may be necessary to encourage traders to enter these markets. Backloading opportunities may also exist for traders to make use of empty trucks returning from carrying emergency supplies into the affected areas;
- compensating local authorities for temporary reductions/suspensions of local fees and levies.

9. Ensuring ongoing support

Having identified the support required, it is important that the agency ensures it has the necessary resources for the duration of the activity. Support should be flexible enough to respond to changing circumstances, such as when the condition of the animals deteriorates to a level where they cannot be sold.

Technical standards
Livestock offtake

10. Monitoring

It is important that agencies keep qualitative and quantitative records of the operation for evaluation, impact assessment, and documentation of best practices. See the livestock offtake monitoring and evaluation indicators in *Appendix 8.2: Examples of monitoring and evaluation indicators for livestock offtake*.

> **Standard 4: Slaughter livestock offtake for consumption**
>
> The intervention salvages value from crisis-affected livestock to provide cash, meat and employment to affected communities.

Key actions

- Involve the affected communities *(see Guidance note 1)*.
- Determine purchase sites and market dates, and publicise them through community participation *(see Guidance note 2)*.
- Agree on purchase prices and payment methods for each species and class of animal *(see Guidance notes 3, 4 and 5)*.
- Assess and agree on opportunities for processing hides and skins *(see Guidance note 6)*.
- Agree on criteria for selecting those most affected, and in-kind contributions, and identify the participants *(see Guidance notes 6, 7 and 8)*.
- Agree on criteria for selecting animals for slaughter *(see Guidance note 9)*.
- Agree on criteria for distributing fresh or dried meat *(see Guidance note 10)*.
- Follow local customs concerning slaughter, butchering and preservation methods, and observe animal welfare standards *(see Guidance note 11)*.
- Assess and act upon public health risks associated with animal slaughter *(see Guidance note 12)*.
- Safely dispose of carcasses unfit for human consumption *(see Guidance note 13)*.
- Provide appropriate equipment and resources for livestock offtake operations *(see Guidance note 14)*.

Technical standards
Livestock offtake

Guidance notes

1. Community involvement

Coordination arrangements from earlier livestock offtake interventions may be resurrected or new groups established (see also *Standards 1* and *2*) to assist in planning and implementation. Details to be determined will include:

- selection criteria for different groups of people affected;
- selection criteria for animals to be purchased for slaughter;
- sites and dates of temporary market;
- whether vouchers, mobile money or pre-paid cards should be used instead of physical cash (see also *Guidance note 4*);
- suitable slaughter sites;
- criteria for when to distribute fresh or dried meat.

2. Purchase sites and dates

Temporary markets should be as close as possible to the affected communities to avoid excessive trekking of already weakened animals. Market days should be fixed in advance and well publicised. Markets should also be scheduled so as to allow adequate time for agency staff to rotate between the sites. The availability of basic infrastructure (holding areas, water, feed, etc.) and services (veterinary inspectors, agency staff, etc.) should be ensured.

3. Purchase price

The purchase price for the different species and types of animals needs to be agreed with and publicised in the affected communities. Coordination with other agencies operating similar schemes in adjacent areas is essential to avoid competition and confusion. Actual market prices, if available, should be monitored and the intervention price (what the agency pays) reviewed and, if necessary, adjusted accordingly. The intervention price is often higher than the actual market price, which may still be too low to benefit prospective sellers. However, if the intervention price is set too high, it may benefit only a small number of sellers and destabilise an already fragile market.

4. Mobile money and vouchers

Agencies should consider using mobile money transfers or vouchers rather than carrying cash in high-risk areas. The spread of mobile technology in the last decade has made it safer, easier and faster to transfer money even to remote areas. Vouchers or pre-paid cards that are usually redeemed later are also preferred to carrying cash.

Technical standards
Livestock offtake

5. Procurement

Agencies may purchase animals directly or contract out to local groups or individuals. Contracting out, where possible, is preferable because it is simpler, less costly and supports local institutions. Both the price the agency pays the contractor and the price the contractor pays the producer must be transparent and agreed (see *Process case study: Contract purchase for livestock offtake for slaughter in Kenya*).

6. Selection of recipients

Slaughter livestock offtake for consumption involves different members of affected communities, who need to be identified and selected as recipients. Agreement over who owns and benefits from the hides and skins also needs to be agreed (see *Process case study: Voucher scheme and meat distribution for livestock offtake in Kenya*).

7. Meat distribution

Meat recipients can be individual households, local institutions (schools, hospitals, prisons), or camps for displaced people. Criteria for distribution should include at-risk individuals, such as unaccompanied and separated children, and child-headed households. Meat distributions may be organised through the coordination group, local leaders or in conjunction with an ongoing food supply operation. This latter would have its own selection criteria and distribution networks.

8. In-kind contributions

Most communities benefiting from a slaughter livestock offtake intervention are expected to make some sort of in-kind contribution. The implementing agency needs to negotiate and agree these contributions. They could include taking responsibility for security arrangements and/or contributing labour or materials.

9. Selection of animals for slaughter

As with commercial livestock offtake, a slaughter livestock offtake intervention should give priority to older, non-reproductive stock (mainly surplus males). Young breeding stock should be excluded if possible, to secure the core breeding herd and therefore livelihoods.

10. Fresh versus dried meat

Fresh meat is generally considered preferable by many communities (fresh meat satiates hunger better than dried meat). It is also the simplest option. Because fresh meat is perishable, however, the logistics of distribution limit the number of animals that can be slaughtered at any one time. Drying meat has the advantage of allowing more animals to be slaughtered and the surplus meat to be preserved for later use. Preservation also allows for a more staggered and widespread distribution than is possible with fresh meat, assuming that dried meat is culturally acceptable. It also has the additional advantage of providing extra employment and the opportunity for communities to acquire new skills. Drying meat safely requires additional preparation, hygienic facilities, clean water and suitable storage facilities.

11. Slaughter methods

Killing and butchering animals should be based on local customs and expertise, provided that animal welfare standards are not compromised. Ensuring animals are dispatched humanely and safely requires basic equipment (ropes, pulleys, captive-bolt stun guns, knives and saws, buckets/plastic crates, etc.). It also requires simple slaughter slabs with access to water, fly protection, and the means to collect and dispose of blood and waste material. Sufficient labour must be available to carry out the work, with agencies providing training and supervision if required.

12. Public health risks

Certain animal diseases (zoonoses), such as anthrax and Rift Valley fever, and parasites *(Echinococcus,* hydatid cysts*)* are transmissible to humans, particularly people already stressed by hunger and malnutrition. Implementing agencies should conduct an assessment of the potential risks to public health before proposing slaughter interventions. Ante- and post-mortem inspection by qualified personnel of all animals and carcasses is essential. Any animal or carcass that is unfit for human consumption should be safely disposed of *(see Guidance note 13).* Rotating slaughter sites can help minimise the risk of spreading disease. Meat is highly perishable, and good hygiene is essential to reduce the risk of food-borne disease. Slaughter and butchering in camp settings may require careful planning and the construction of temporary facilities to ensure public health and avoid the spread of disease.

Technical standards
Livestock offtake

13. Disposal of condemned carcasses and slaughter waste

Condemned carcasses and waste water, stomach contents, etc. need to be safely disposed of. This usually involves burying (preferably with lime), burning, or quarantining the carcasses. Waste water and body contents must not contaminate sources of drinking water *(see Chapter 6: Veterinary support)*.

14. Providing necessary equipment/resources

Implementing agencies may need to consider supplying necessary items from outside (such as ropes, knives, buckets, nails, etc.) in areas where slaughter livestock offtake for consumption operations take place. Such items could be crucial to facilitate livestock offtake operations.

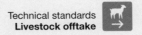

Technical standards
Livestock offtake

Appendices

Appendix 8.1: Assessment checklist for livestock offtake

For livestock offtake in general

- What phase has the emergency reached?
- What is the condition of the livestock being brought to market?
- Is the number of livestock being brought to market increasing?
- What is happening to the price of livestock?
- Which stakeholders are operating in the area?
- Which are the most at-risk communities, households and individuals affected by the emergency?
- How is the crisis impacting different groups, including women and people with disabilities?
- Who could benefit from livestock offtake?
- Can a coordination group be established?
- Have animal welfare standards been considered?
- Is the area secure for the movement of stock and cash?
- What customary and local institutions exist that can facilitate livestock offtake? What roles do they play?

For commercial livestock offtake

- Are traders already operating in the area?
- Is the infrastructure in place to enable livestock offtake?
- Do (temporary) holding grounds exist?
- Is there access for trucks?
- Are feed and water available?
- Are there animal welfare issues regarding trucking livestock?
- Are there any key policy constraints to livestock movement and trade?
- What constraints would hamper access to markets by the most at-risk people and groups?

For slaughter livestock offtake for human consumption

- What slaughter facilities exist?
- What are the local religious and cultural requirements regarding livestock slaughter? Do they compromise accepted animal welfare standards?

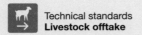
**Technical standards
Livestock offtake**

- What are the local gender roles regarding slaughter, meat preparation, tanning, etc.?
- Which are the most at-risk communities, households and individuals affected by the emergency who could benefit from the slaughter of animals?
- Should temporary market sites be established to reach remote villages?
- Which at-risk groups should be targeted to receive the meat from livestock offtake operations?
- Which individuals could benefit from the employment opportunities that slaughter livestock offtake could provide?
- Can acceptable ante- and post-mortem inspections be undertaken?
- Can a system be established to process hides and skins?

For slaughter for disposal
- Can the hides and skins of condemned carcasses be processed?
- What provisions exist for safe disposal of carcasses?

Appendix 8.2: Examples of monitoring and evaluation indicators for livestock offtake

	Process indicators (measure things happening)	Impact indicators (measure the result of things happening)
Commercial offtake		
Designing the intervention	Number of meetings held with government and traders; range and type of stakeholders participating in meetings	Meeting minutes with an action plan and a clear description of the roles and responsibilities of the different actors
	Number of community-level meetings; number and type of people participating in meetings	Trader preferences for types of livestock for purchase, documented against market demands
		Holding areas clearly defined as needed
		Taxes and other administrative issues agreed with government
		Community-level action plans developed, with agreed prices for livestock, payment mechanisms, and system and schedule for local collection and purchase of livestock
Implementation: livestock purchases	Number of traders involved	Income derived from livestock sales by household
	Number and type of livestock purchased by household and area	Uses of income derived from livestock sales (e.g. to buy food; buy livestock feed; relocate animals; buy medicines, pay school fees)
		Influence on policy

continued over

Technical standards
Livestock offtake

	Process indicators (measure things happening)	Impact indicators (measure the result of things happening)
Slaughter livestock offtake for consumption		
Designing the intervention	Number of community-level meetings; number, position and gender representation of people participating in meetings Formation of community-level livestock offtake coordination group in each target location	Meeting minutes with clear description of the roles and responsibilities of the different actors Terms of reference for livestock offtake coordination group agreed Action plans developed with agreement on: Setting temporary market days, and frequency/duration of intervention Selection criteria for affected communities Types of livestock for purchase together with prices and payment mechanisms Amount of meat to be distributed System for local collection and purchase of livestock, with timings Hire and payment of community members involved in slaughter, meat preparation, handling skins, etc.
Implementation: slaughter and meat distribution	Number of participating households and people Number and type of livestock purchased by household and area Amount of meat distributed per household Number of local people hired for temporary work	People selling livestock – income derived from livestock sales by household and uses of income People hired for temporary work – income received and uses of income Improved nutritional status of meat recipients

References and further reading

Abebe, D., Cullis, A., Catley, A., Aklilu, Y., Mekonnen, G. and Ghebrechirstos, Y. (2008) 'Impact of a commercial de-stocking relief intervention in Moyale district, southern Ethiopia', *Disasters* 32(2): 167–186, http://dx.doi.org/10.1111/j.1467-7717.2007.01034.x

Aklilu, Y. and Wekesa, M. (2002) *Drought, Livestock and Livelihoods: Lessons from the 1999–2001 Emergency Response in the Pastoral Sector in Kenya*, Humanitarian Practice Network Paper 40, Overseas Development Institute (ODI), London, https://www.unisdr.org/files/1855_VL102122.pdf

Catley, A. (ed.) (2007) *Impact Assessment of Livelihoods-based Drought Interventions in Moyale and Dire Woredas, Ethiopia*, Pastoralists Livelihoods Initiative, Feinstein International Center, Tufts University, Medford, MA, together with CARE, Save the Children USA, and USAID-Ethiopia, https://fic.tufts.edu/assets/IMPACT1-2.pdf

Catley, A. and Cullis, A. (2012) 'Money to burn? Comparing the costs and benefits of drought responses in pastoralist areas of Ethiopia', *Journal of Humanitarian Assistance*, 24 April 2012, https://www.livestock-emergency.net/wp-content/uploads/2020/05/Catley-and-Cullis-2012.pdf

FAO (2001) *Guidelines for humane handling, transport and slaughter of livestock*, https://www.fao.org/documents/card/en/c/466c5e0e-aa6d-5cdb-8476-43245054c3bf/

FAO (2016) *Livestock-related interventions during emergencies – The how-to-do-it manual*. Edited by Philippe Ankers, Suzan Bishop, Simon Mack and Klaas Dietze. FAO Animal Production and Health Manual No. 18. Rome, https://www.fao.org/3/i5904e/i5904e.pdf (See Chapter 4)

FAO (2020) *Effective disposal of animal carcasses and contaminated material on small to medium-sized farms*, FAO Animal Production and Health/Guidelines 23, https://www.fao.org/publications/card/en/c/CB2464EN/

Morton, J. and Barton D. (2002) 'Destocking as a drought-mitigation strategy: clarifying rationales and answering critiques', *Disasters* 26(3): 213–228, http://dx.doi.org/10.1111/1467-7717.00201

Toulmin, C. (1994) 'Tracking through drought: options for destocking and restocking', in I. Scoones (ed.), *Living with Uncertainty: New Directions in Pastoral Development in Africa*, pp. 95–115, Intermediate Technology Publications, London.

Turner, M.D. and Williams, T.O. (2002) 'Livestock market dynamics and local vulnerabilities in the Sahel', *World Development* 30(4): 683–705, http://dx.doi.org/10.1016/S0305-750X(01)00133-4

See also case studies for livestock offtake in emergencies at: https://www.livestock-emergency.net/resources/case-studies/

Technical standards
Provision of livestock

Chapter 9: Technical standards for the provision of livestock

323 **Introduction**

325 **Options for the provision of livestock**

330 **Timing of interventions**

331 **Links to the other LEGS chapters and other HSP standards**

331 **LEGS Principles and other issues to consider**

338 **Decision tree for provision of livestock options**

341 **The standards**

352 **Appendix 9.1: Assessment checklist for provision of livestock**

354 **Appendix 9.2: Examples of monitoring and evaluation indicators for the provision of livestock**

358 **Appendix 9.3: Discussion on minimum viable herd size**

360 **References and further reading**

Photo © Kertu/iStock

Technical standards
Provision of livestock

Chapter 9: Technical standards for the provision of livestock

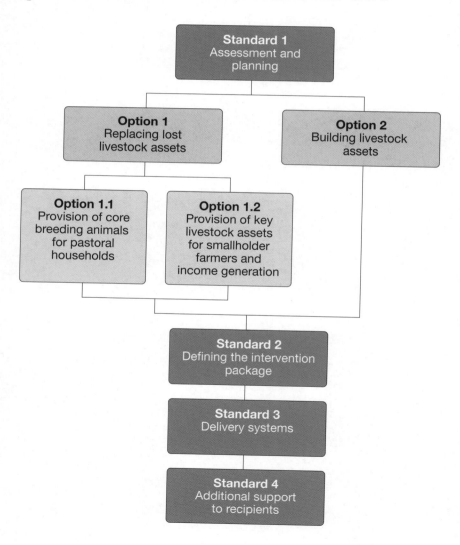

Technical standards
Provision of livestock

Introduction

When emergencies result in substantial loss of livestock, the provision of livestock is a valuable approach to rebuilding people's financial livelihood assets and providing high-quality livestock-derived foods.

Different emergencies result in the loss of different numbers of livestock. For example, a small flood may result in no livestock losses, since animals can be moved to higher ground (and floodwater may in fact increase the availability of forage and pasture in the medium term). Large-scale flooding, on the other hand, may result in the loss of thousands of livestock. Similarly, a severe drought or a strong earthquake may result in the loss of large numbers of animals. Volcanic ash may cause livestock death through starvation if fodder and feed resources are not made available.

Communities that have lost key livestock or a significant number of their core breeding animals in an emergency will benefit from livestock provided in the recovery phase. This ensures their livelihoods can be supported and potentially rebuilt. An intervention is regarded as successful when the animals provided survive, increase in number, are productive and contribute to the livelihoods of the crisis-affected communities. In pastoral and smallholder farmer contexts, the provision of livestock has an immediate positive outcome. This is because productive animals provide milk and eggs and help improve child and household nutrition and health. Transport animals, meanwhile, facilitate livelihood activities and/or generate income.

This chapter presents information on the provision of livestock as a livestock emergency response, together with technical options for the provision of livestock and the associated benefits and challenges of each. Information is also available in Chapter 9: Provision of Livestock in *FAO* (2016). For each technical option, LEGS provides information through standards, key actions and guidance notes. Checklists for assessment, as well as monitoring and evaluation indicators, are presented as appendices at the end of this chapter. Further reading is also provided. Case studies are presented on the LEGS website (see https://www.livestock-emergency.net/case-studies).

Links to the LEGS livelihoods objectives

The provision of livestock in the recovery phase of an emergency supports the third LEGS livelihoods objective: **to support crisis-affected communities to rebuild key livestock-related assets**.

Technical standards
Provision of livestock

The provision of livestock should always meet all five of the animal welfare domains (see Chapter 1: Introduction to LEGS). This is likely to result in improved survival rates and higher levels of productivity, and therefore contribute better to positive livelihood outcomes.

The importance of the provision of livestock in emergency response

Replacing livestock after emergencies, such as drought, flood, livestock disease, volcanic eruptions, earthquakes, and conflict – including in some areas, militarised livestock theft – helps rebuild the livelihoods of affected households. Wherever possible, however, it is preferable to reduce livestock losses during the emergency phase itself. This can be achieved by using LEGS technical interventions such as the provision of livestock feed and water, and veterinary support (see Chapters 4, 5 and 6).

During and immediately after humanitarian emergencies, the focus is on the provision of food, water, shelter and health for human populations. Done well to Sphere standards, such assistance reduces the need for crisis-affected communities to sell remaining livestock to meet basic household needs. As basic human needs are met, additional important assistance can be provided to protect their remaining animals by delivering feed, water, animal health, and shelter support (see Chapters 4–7). In contrast, interventions focused on replacing livestock are typically delayed until the recovery phase of an emergency. By this phase, both crisis-affected communities and agency staff have the time and capacity to assess the appropriateness of a provision of livestock intervention. It is important that these assessments are not hurried and that interventions also fully and appropriately provide for associated livestock support needs, such as animal health, feed, water and shelter. Agency technical capacity should also not be stretched too thinly.

The provision of livestock is a technically challenging intervention. Tasks include:

- assessing and agreeing the types and numbers of animals involved;
- the selection of individual animals;
- the choice of recipients (and inevitably the tough decision to exclude other recipients); and
- how to address subsequent issues such as livestock barrenness, loss or theft.

The provision of livestock is also costly. This is particularly true when an emergency has resulted in the loss of large numbers of livestock, and local livestock prices are inflated. Costs are further increased when recipients continue to need other forms of support for their remaining animals. While cost concerns are important, the cost of providing ongoing humanitarian support to crisis-affected communities might be higher in the long term than the cost of rebuilding their livelihoods. Providing livestock may therefore contribute to a reduction in the overall costs of supporting a humanitarian intervention.

Given the complexities and costs of the provision of livestock, it is essential that implementing agencies have the necessary specialist livestock and social science staff. If not, it may be better for them to partner with a specialist livestock agency and, over time, build the necessary skills and capacities. Whether experienced or not, agencies need to be able to demonstrate that they understand local livestock economies. This is important for them if they are to win the confidence of local livestock keepers, customary institutions (in pastoral areas), local farmer organisations and the local administration. It is also important that agencies work well with all wealth groups, women, and marginalised groups who might not normally be integrated in smallholder farmer and pastoral organisations.

Agencies with long-term development experience in an area affected by the emergency are often well placed to lead an intervention. This is especially true if they are familiar with local livestock production systems and the associated social systems that underpin them. However, this does not necessarily avoid basic errors; exotic livestock, for example, are known to have been distributed in areas of limited agro-ecological potential. The limited capacity of exotic livestock to adapt to the conditions results in their low productivity and high mortality rates.

Agencies providing replacement animals may want to consider aligning the intervention with longer-term livestock development initiatives, and interventions that support market-based approaches. For example, it may be possible for them to support recipients to explore and use local fodder and livestock markets. Or they might support recipients' access to private animal healthcare services, including through the use of multi-sectoral cash grants.

Options for the provision of livestock

There are two main options for providing livestock. These depend on the nature of the emergency, the local livestock production system, livelihood strategies and the operational opportunities and challenges. The two options are 'replacing lost livestock' and 'building livestock assets'.

Technical standards
Provision of livestock

To be successful, it is important that informed decisions for both technical options are made based on the following:

- the selection of livestock, specifically the age, sex, number, and type of animals as well as their health status and price;
- the number of potential recipients and their required capacities, including labour, experience, skills and interests; as well as their access to essential productive resources and services – feed, water, shelter and animal healthcare.

The standards presented in this chapter offer guidance that is relevant to both options.

Option 1: Replacing lost livestock assets

Under this technical option, agencies may provide livestock to different livelihood groups under two different sub-options:

Sub-option 1: Provision of core breeding animals for pastoral households

Pastoral (and agro-pastoral) communities inhabit the world's grasslands, stretching from China and Mongolia, through Central Asia and the Middle East, to the Sudano-Sahelian zone and southern Africa. Other pastoralists live in Central and South America, Europe, North America and Australasia. All depend heavily on livestock as a source of food and income, and livestock are typically at the centre of their social networks. Emergencies that decimate flocks and herds therefore have very damaging impacts on pastoral livelihoods and may result in psychosocial stress or destitution.

Pastoralism differs from area to area, as some rely on a single animal type (for example, sheep, cattle or reindeer), while others keep mixed herds of sheep, goats, cattle/yaks and camels. Given this diversity, it is important that local knowledge drives operational choices and the approaches that inform interventions. For example, camels may be highly prized by some pastoral communities, which might suggest that camels should be central to an intervention. However, local knowledge also adds that small ruminants, such as sheep and goats, are valued because they can be traded up for camels as they increase in number. Using local knowledge in this way can help reduce intervention costs. Local informants should play a central role in determining the number of animals that constitute a minimum 'core breeding' flock/herd (see *Appendix 9.3: Discussion on minimum viable herd size* at the end of this chapter).

The redistribution of livestock through marriage, payment of fines and friendship, and the associated building of social networks, is common in almost all pastoral societies. It makes good sense for agencies to use this knowledge, and the mechanisms and institutions that support them. For example, local knowledge and institutions can help them identify when and to whom livestock should be provided (see also LEGS *Principle 2: Community participation*).

It is important to support only those pastoralists who have a real interest in returning to the pastoral sector and who retain good links with other pastoralists. This is because pastoral households cannot function in isolation from others and need to cooperate in the herding and watering of livestock. It is for these reasons that pastoralists typically herd communally in groups of households. Working together they share labour, rotate breeding males, and can safely access more remote grazing areas. Pastoralists who no longer have such social links are unlikely to make a sustainable return to pastoralism.

Sub-option 2: Provision of key livestock assets for smallholder farmers and for income generation

Poor and smallholder farmers, and people living in peri-urban and urban areas, may own relatively few animals – certainly compared with pastoral households. Wealthier smallholder farmers may own several dairy cows, a pair of ploughing oxen, sheep and goats, a horse or mule, some pigs, and poultry flocks (chickens, ducks and turkeys). In contrast, very poor smallholder farmers may own just a few chickens.

Households who own a small herd of dairy cows, small flocks of poultry or some working animals will be highly dependent on those animals. They will produce and sell dairy products and eggs, or hire out their working animals for the transport of produce and goods. In Ethiopia, donkeys and mules have a long history of use in the transport of coffee, while working animals in Nepal transport goods to remote mountain communities. Working animals are also used to deliver food aid and other forms of humanitarian assistance in times of crisis.

Peri-urban and urban livestock, even in these small numbers, represent an important livelihood source. Their loss in an emergency may result in acute suffering, including loss of food and income. Over time, such losses may also result in more limited use, or even complete loss, of valuable knowledge, skills, and experience. For affected communities, replacing animals lost to an emergency without external support may also contribute further to the deepening of the impact of the emergency. This is because they must use scarce resources and assets to secure the necessary funds for replacement.

Option 2: Building livestock assets

Under this technical option, agencies may provide livestock to help improve livelihood outcomes after an emergency for people who have not previously owned livestock before. Livestock may improve livelihoods in the following ways:

- by increasing access to milk and eggs and improving nutritional outcomes;
- as an additional source of income generation;
- as a form of saving that can later be mobilised in times of need, including for the payment of school or medical fees through the sale of animals (if more are born than sold).

Through the provision of livestock, including to internally displaced persons (IDPs) and refugees, livelihood diversification is supported. The intervention may also help make better use of naturally occurring forage and feed resources. In so doing, it may convert local resources, which cannot be consumed by human populations, into high-quality animal-source protein. Livestock may also provide power – through transport and ploughing – and therefore contribute income and create favourable conditions for planting crops. Recipients can also use manure from livestock as a source of fertiliser, as a household fuel, for construction or for sale.

While there are some known successes in the provision of livestock to people without prior experience of livestock keeping, there are also failures. This is because keeping livestock has its challenges: time, knowledge, experience and passion for livestock are all requirements for keeping healthy and productive animals. Where there is strong interest, however, it may be appropriate for livestock provision interventions to start small, with poultry. This is because poultry have relatively short production cycles and can almost immediately contribute to household nutrition and food security, or provide income through the sale of eggs.

In these cases, to ensure positive outcomes, it is vital to organise appropriate training in animal nutrition, husbandry, health and breeding. Where possible, specialist support should also continue to be offered for a minimum of two years after the intervention. In this way, it may be possible to provide recipients with the full range of knowledge and skills required to keep livestock successfully.

The benefits, challenges and implications of these options and sub-options are summarised in *Table 9.1*.

Table 9.1: Benefits and challenges of livestock provision options

Benefits	Challenges	Implications
Option 1: Replacing lost livestock assets		
Sub-option 1: Replacing core breeding animals for pastoralists		
Helps replace lost livestock	Unit cost of the intervention is high, and monitoring, evaluation and impact assessment require commitment and resources	Specific to pastoral communities
Contributes to food and nutrition security		Recipients must retain social status and access to pasture and water
Contributes to resilience strengthening		
Supports customary livestock exchanges and herd reconstitution	Requires considerable technical and social science support	Recipients may require other livelihood support (food and cash) while livestock multiply
May eventually support other livelihoods (traders, transporters, processors, etc.)		Livestock require other forms of support (veterinary, feed, shelter, etc.)
Sub-option 2. Provision of key livestock for smallholder farmers and for income generation		
Helps replace lost livestock	Unit cost of the intervention is high, and monitoring, evaluation and impact assessment require commitment and resources	Recipients may require other livelihood support (food and cash) while livestock multiply
Contributes to food and nutrition security		
Contributes to income generation		Livestock require other forms of support (veterinary, feed, shelter, etc.)

continued over

Technical standards
Provision of livestock

Benefits	Challenges	Implications
Option 2: Building livestock assets		
Provides valuable new assets Contributes to food and nutrition security Contributes to income generation	Unit cost of intervention is high, and monitoring, evaluation and impact assessment require commitment and resources Introduction of livestock needs to be supported by training	Training and mentoring support should be continued over two years

Timing of interventions

The costs and challenges associated with the delivery of high-quality livestock provision interventions require that agencies make every effort to ensure the best possible outcomes. This includes delaying interventions until the recovery phase of the disaster risk management cycle (see Table 9.2). In this way, they are more likely to ensure minimal competition with the provision of critical life-saving support in the emergency phase. The premature sale or consumption of distributed livestock to meet urgent needs will also be less likely. And support for the livestock sector can focus on the provision of veterinary support, feed, water and shelter.

Agencies should further delay the timing after a complex emergency, as there are potential ongoing dangers to both livestock and livestock owners from armed groups. This may be the case even after peace agreements have been signed.

Table 9.2: Possible timing of livestock provision

Options	Rapid-onset emergency		
	Immediately after	Early recovery	Recovery
1. Replacing lost livestock assets	—	—	✓
2. Building livestock assets	—	—	✓

Options	Slow-onset emergency			
	Alert	Alarm	Emergency	Recovery
1. Replacing lost livestock assets	—	—	—	✓
2. Building livestock assets	—	—	—	✓

Links to the other LEGS chapters and other HSP standards

The provision of livestock should be linked to and coordinated with other LEGS interventions, including the provision of feed, water, veterinary support and shelter *(see LEGS Chapters 4, 5, 6 and 7).*

Recipient households are also likely to require other forms of assistance to meet their basic needs, at least until the livestock provided can make a full contribution to livelihoods. For example, a pastoral (or agro-pastoral) recipient who receives 20 or 30 core breeding animals will not achieve household food security for several years. They will therefore require food and income during the flock/herd rebuilding period (see *Standard 4: Additional support* below). Without such assistance, they may be forced to sell the animals that they have received to meet their basic food and income needs. Agencies should consult the 'Minimum standards in food security and nutrition', as well as the 'Minimum standards for shelter and settlement' in the Sphere Handbook. By doing so, they can ensure that recipient households receive all forms of necessary support during the recovery period.

Technical standards
Provision of livestock

LEGS Principles and other issues to consider

The provision of livestock, either by replacing lost livestock assets or building livestock assets as a new livelihood activity, can play an important role in strengthening and rebuilding livelihoods affected by emergencies. LEGS Principles can help inform approaches to such interventions, as indicated in *Table 9.3*.

Table 9.3: LEGS Principles and how they relate to provision of livestock

LEGS Principle	Examples of how the principles are relevant in provision of livestock interventions
1. Supporting livelihoods-based programming	The provision of livestock strengthens and rebuilds livelihoods. This is either through replacing core breeding animals in pastoral herds, or replacing the livestock that smallholder farmers depend on for income generation. Livestock provision may also be used as a new livelihood activity.
	The use of market approaches also supports the livelihoods of local service providers – traders, transporters, livestock input suppliers, animal health providers, etc.
2. Ensuring community participation	Better outcomes are achieved when affected communities participate in all aspects of a provision of livestock intervention: in other words, assessment, design, implementation, as well as monitoring and evaluation. Specifically, affected communities should be involved in the feasibility assessments and selection of type and number of livestock to be distributed; the selection of recipients; the procurement and distribution of livestock; and the monitoring and evaluation of the intervention.
3. Responding to climate change and protecting the environment	Environmental impact assessments prior to providing livestock should assess current environmental resilience and the ongoing availability of natural resources (grazing and water) that will be required for the new livestock. Issues specific to urban and peri-urban areas (including animal waste) should be considered where appropriate.
	Livestock selected should ideally be local breeds adapted to the local agro-ecological conditions and so be able to survive and be productive.

continued over

Technical standards
Provision of livestock

LEGS Principle	Examples of how the principles are relevant in provision of livestock interventions
4. Supporting preparedness and early action	Communities due to receive livestock for the first time need to be prepared and to undertake appropriate animal husbandry skills training.
	Detailed preparatory assessments are necessary to ensure the availability of feed, water and shelter, as well as the commitment levels of proposed recipients.
	Strategies to ensure preparedness for future emergencies, such as disaster risk reduction techniques, should be made available to recipients.
5. Ensuring coordinated responses	Coordination with other agencies involved in the provision of livestock is necessary to ensure conflicting approaches are not used.
	Coordination of transportation arrangements is necessary to ensure animal welfare requirements are met throughout their journey.
	Coordinated agency support to recipients for the provision of ongoing humanitarian assistance – food, income, etc. – may be necessary for several years, until livestock become productive.
	Coordination with complementary livestock interventions – veterinary support, feed, water, etc. – is also needed to ensure the survival of the livestock provided.

continued over

Technical standards
Provision of livestock

LEGS Principle	Examples of how the principles are relevant in provision of livestock interventions
6. Supporting gender-sensitive programming	For the intervention to be effective, men and women of different wealth categories should be asked to participate fully in the design, implementation and management of livestock provision interventions, as well as in the monitoring, evaluation, accountability and learning (MEAL). Women should be given sufficient opportunity to be able to express their particular interest.
	Where women have adequate labour and time, and have expressed interest in being recipients, their priorities should be assessed in terms of species selection.
	Where an intervention can include female-headed households, agencies should offer additional ongoing training and mentoring support as required.
7. Supporting local ownership	Local customary/administrative leaders (male and female) should play a central leadership role in the intervention.
	Long-standing and customary livestock redistribution mechanisms should be supported and not undermined by the intervention.
	Interventions should make use of local skills, knowledge and livestock management capacities.
8. Committing to MEAL	It is important for agencies to monitor and evaluate the impact of the provision of livestock. They should focus in particular on the survival and reproduction rates of livestock that they have provided in the months and years after the emergency. This is to determine the ultimate impact of the intervention on livelihoods, and to generate learning and guidance for future interventions.

Security and conflict

It is important to give full consideration to protection issues as, in areas of conflict, the provision of livestock may attract the interest of armed groups and criminals. This is a particular problem where interventions are provided to female-headed households and minority or marginalised groups. In some conflict areas, livestock raiding has become militarised, and groups of well-armed men travel long distances to raid livestock. If there is any concern that livestock, and therefore livestock keepers, are at risk from raiding, an intervention should not be carried out. In all instances Sphere Protection Principle 1, 'Avoid exposing people to further harm', should be considered (see *Chapter 1: Introduction to LEGS*).

Similarly, in areas where pastoral and sedentary smallholder farmers are in conflict, agencies need to exercise caution before providing one or other group with additional livestock. The selection of different species may help reduce vulnerability to theft. For example, goats may be less attractive to raiders than cattle as they are more difficult to drive away at speed. Agencies providing livestock also need to ensure that purchased animals, or animals to be purchased using cash or vouchers they have distributed, are not themselves stolen. If there is any risk in this regard, the intervention should be stopped.

Selection of recipients

Reference has been made to the need to select recipients with care. For example, in pastoral areas it is of critical importance that potential recipients have some animals or can confirm their recent destitution is the result of an emergency. They also need to continue to retain strong links with their pastoral community. Only through such strong links can recipients integrate back into a functional herding unit and become part of the wider pastoral community. In this way, they can also receive livestock gifts and loans (for example, for breeding purposes), and ultimately make gifts and loans themselves as their animal numbers increase. Such transactions may well prove central to a household's longer-term viability (see *Process case study: Supporting traditional livestock distribution as a drought-preparedness strategy in Niger*).

In the case of smallholder farmers, it is important that recipient households have the labour and capacity to manage and provide for the animals they receive. This will include them providing appropriate healthcare, suitable grazing and feed and water, shelter/housing, and healthy stock for breeding. Where agencies provide animals to households with little or no previous

Technical standards
Provision of livestock

history of owning and managing livestock, they need to give special care to the selection of recipients. Only by doing this, will it be possible to screen out those who lack the time, capacity and real interest required to maintain healthy and productive animals.

Typically, the families that can make the best from the provision of livestock are neither the poorest nor the most at risk, who often live in roadside locations or on the peripheries of market towns. To identify the households that the provision of livestock may best help requires involving customary and clan leaders. However, care has to be taken to ensure that those who are screened out are not consistently female-headed households or those from minority groups *(see LEGS Principle 6: Gender-sensitive programming and Principle 7: Local ownership)*.

Where interventions are making a special effort to benefit very poor or disabled people, the provision of just two or three hens may be as much as can be managed. These should be provided together with appropriate housing and access to feed and healthcare through to the point of laying. Whenever possible, it is important to complement interventions with gender-sensitive programming that reflects species prioritisation and appropriate technical advice (for example, subsidised poultry healthcare) for the following 12 months.

Disease transmission

The transmission of disease from livestock to human populations typically occurs in more industrial livestock production systems. However, it can occur in any setting where livestock and people live close to each other. Examples include pastoralists and smallholder farmers whose animals may shelter in part of the home at night. For instance, chickens may roost in the roof, or sheep and goats, cattle, and working animals may be corralled under a sleeping platform for safety. In some towns and cities, and in some IDP/refugee camps (especially those that include pastoralists and smallholder farmers), large numbers of livestock may live near human populations *(see Chapter 7: Shelter)*. In each of these settings, interventions should safeguard animal and human populations through providing access to healthcare and the regulation of minimum sanitary and hygiene conditions.

Use of cash and voucher assistance, and livestock fairs

The approaches used to provide livestock have changed over time. In the 1990s, most agencies contracted local traders to purchase breeding animals that were shared among recipients. In some cases, where several hundred sheep and goats were purchased and distributed in pastoral settings, each recipient would pick an animal in turn, until all the animals were distributed. In this way, it was stated that both better and less good animals were more equally shared.

Today, cash and voucher assistance (CVA) is generally preferred, as recipients are afforded better choice, and full use is made of local markets (see *Chapter 3* on CVA). Under this approach, following discussions with the local market manager, recipient households are given either cash or vouchers with which to purchase livestock. (See *Impact case study: Herd replacement using cash vouchers in Kenya* for an example of cash transfers used for replacing livestock assets).

Local markets may also agree to hold special livestock fairs where, following an announcement, local pastoralists or livestock keepers bring surplus animals of an agreed age, sex, breed and type for sale. On the livestock fair day, agencies provide recipients with cash or vouchers to purchase agreed types and numbers of animals. For example, an intervention may focus on the provision of two- to three-year-old heifers or on one-year-old female sheep and goats. Whichever is chosen, recipients understand the types and numbers of animals that they can purchase. When the transactions are concluded, the agencies repay the vouchers in local currency to the sellers. Livestock fairs can be suitable for all livestock provision options. Markets and livestock fairs also encourage information and knowledge sharing (see *Impact case study: Livestock fairs in Niger)*.

While the use of cash and vouchers may help improve choice and support local market systems, including through hosted fair days, it is important to remember that few livestock keepers sell their best animals. It may therefore take some years of careful breeding before recipients build back the quality of genetics that they owned before the onset of the emergency.

Camps

The provision of livestock to people living in IDP and refugee camps can only be considered after discussion and agreement with the camp management. This is because there are implications for health and sanitation and, according to the animal type, for the use of local water and grazing resources. Where camp-based animals compete for resources, there is

Technical standards
Provision of livestock

always the potential for conflict with local livestock keepers. It is essential, therefore, to hold discussions with host community representatives if sheep, goats and particularly cattle, are being considered as part of an intervention. These discussions are increasingly important, as while camps are temporary arrangements, many have lasted for years. Poorly planned interventions can become the source of long-running friction *(see Chapter 7: Shelter)*.

Poultry are often a first choice for agencies supporting interventions in IDP and refugee camps, as the impact on other residents is limited. Poultry scavenge in the area immediately around the recipient's tent/home, and are only a problem if people in the area are trying to crop vegetables and flowers. Where crops are being grown, the intervention should only proceed if poultry are fenced. Such fencing also offers health benefits for children through the avoidance of interaction with animal faeces.

Where fencing is problematic, recipients may be moved to the margins of the camp. This may afford livestock the opportunities to roam and forage more naturally outside the camp. Whatever compromises are agreed, it is important that livestock numbers are not allowed to grow beyond agreed levels for each household or for each animal type.

Decision tree for provision of livestock options

The decision tree *(Figure 9.1)* summarises some of the key questions to consider in determining which may be the most feasible and appropriate option for a provision of livestock intervention. The standards, key actions and guidance notes that follow provide more information for detailed planning.

Technical standards
Provision of livestock

Figure 9.1: Decision tree for the provision of livestock

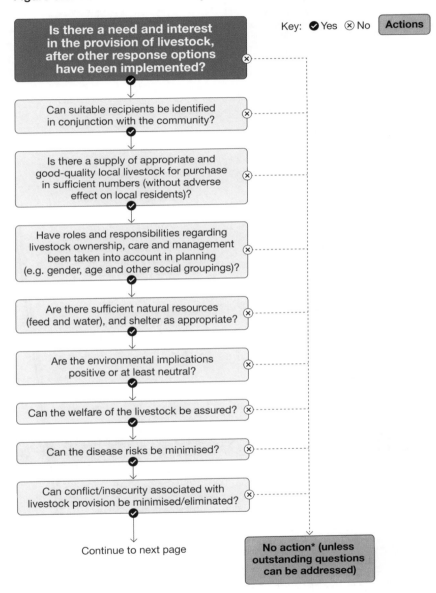

*The result **'No action'** does not necessarily mean that no intervention should take place, but rather that further training or capacity strengthening may be required in order to be able to answer 'yes' to the key questions.

Technical standards
Provision of livestock

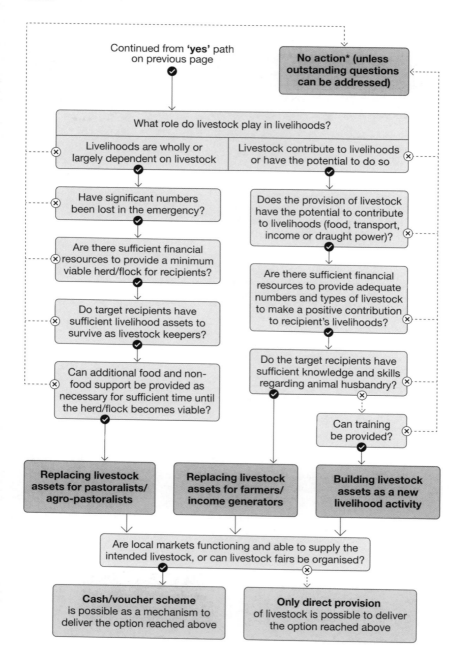

Technical standards
Provision of livestock

The standards

Standard 1: Assessment and planning

Assessments for provision of livestock interventions provide a comprehensive analysis of the role livestock play in different livelihoods, the potential impact, and mitigation measures.

Key actions
- Consider the role of livestock in local livelihoods *(see Guidance note 1)*.
- Assess the likely impact of the intervention on the local livestock economy as well as its animal welfare, environmental and security impacts *(see Guidance note 2)*.
- Where providing animals is considered for people with no livestock-keeping experience, carefully assess the planned recipients' access to key resources. These include feed, water, shelter and animal health services, including labour and social networks *(see Guidance note 3)*.
- Ensure there is a full and detailed assessment of all cost implications *(see Guidance note 4)*.

Guidance notes
1. Livestock in livelihoods
Provision of livestock interventions should be informed by a thorough understanding of the role livestock currently play in the livelihoods of intended recipients in 'normal' times. Such an assessment should include details of animal types and numbers, livestock management systems, and livestock keepers' skills, experience and knowledge.

Providing livestock to pastoral and smallholder farmers can only succeed where recipients retain access to productive assets – land, water, shelter, animal health services, and markets, in other words. They should also still have access to the labour and social networks they need in order to keep livestock successfully. The initial assessment checklist for the role of livestock in livelihoods and livestock management can be found in *Chapter 3: Emergency response planning*; see also *Appendix 9.1: Assessment checklist for the provision of livestock*.

Technical standards
Provision of livestock

2. Impact and mitigating measures

The purchase of livestock, and even the announcement of a provision of livestock intervention, may affect market prices. Such increases may concern traders and wealthier households, who themselves may be engaged in rebuilding their own livestock assets after an emergency. Yet price increases often benefit local producers, who typically sell more animals than they buy. Price increases alone should therefore not prevent the intervention proceeding. To avoid antagonism, however, it may be helpful for agencies to stagger purchases and therefore reduce the impact on the market (see also FAO (2016, p. 146) on risks and mitigation options in restocking).

The purchase of large numbers of animals from outside the area can have negative environmental impacts when the numbers of animals exceed the local environmental limits. Impacts are, however, typically minimal when implementing agencies are purchasing animals locally or when they distribute fewer animals compared with the number lost in an emergency. However, to ensure that full consideration is given to environmental protection, it is advisable to conduct a rapid environmental impact assessment (see also LEGS *Principle 3: Climate change and the environment)*.

Similar care and thorough assessments are required where households living in IDP and refugee camps are potential recipients of livestock. This is because there are wider health, sanitation and security issues.

Pre-intervention assessments should also analyse animal health, welfare and associated disease risks. These should be included in particular where it is anticipated that animals will be trekked or transported for long distances ahead of a market day, held for a day or more in a holding ground or market and then trekked or transported to the recipient's homestead. Throughout this time, it is important that animals have adequate access to water, feed, and shelter (in hot/cold environments), and to adequate periods of rest (see the 'five domains' in *Chapter 1: Introduction to LEGS*).

Finally, the assessment should include the possible security implications of providing livestock, including whether recipient households may become a target for theft or organised raiding. The grazing and watering of livestock may trigger increased conflict when resources are in short supply or when animals are distributed when crops are growing. The provision of livestock should not proceed if there is a reasonable threat that insecurity, theft or conflict will increase.

3. The provision of livestock to non-livestock keepers

Where livestock do not form part of a local livelihood strategy, agencies need to complete a detailed assessment before considering an intervention. Livestock require access to clean water, good food, shelter and protection. If potential recipients cannot ensure the sustained delivery of these essentials, then an intervention should not be considered. Even where these essentials are accessible, the daily workload associated with good livestock husbandry requires significant amounts of labour, knowledge, skills and commitment. An assessment should be made of the likely impact of providing livestock on the men, women, boys and girls in the recipient households, including its potential impact on child labour. Local markets also need to be assessed for potential income opportunities from the sale of products and eventually of livestock. If there is doubt, then an intervention should not be considered. (See *Process case study*: *Livestock distribution and feed supply following an earthquake in Pakistan*. See also *FAO* (2016, p. 146) on risks and mitigation options in restocking.) Where these essentials are met, a thorough training needs assessment should still be carried out to tailor training to the specific needs of recipients in the planned intervention area.

4. Benefit-cost analysis

Recognising the financial and operational costs involved in a provision of livestock intervention, agencies should only consider implementing such an intervention after a thorough benefit-cost assessment.

They should also make an assessment of alternatives. For example, in emergency-affected areas where small numbers of livestock remain, it may be possible for them to support accelerated flock/herd rebuilding at a fraction of the cost of providing livestock. This could be through support to animal health and the occasional use of high-quality, supplementary feed (see *Chapter 6: Veterinary support* and *Chapter 4: Livestock feed*).

Possible policy constraints may also impact the cost of an intervention. These may be external (concerning the purchase or movement of livestock, including market, quarantine, and transport taxes) or internal (for example, the purchasing protocols of the agency involved). These constraints might delay progress and therefore increase costs.

Technical standards
Provision of livestock

Standard 2: Defining the intervention package

Appropriate livestock types and breeds are distributed in adequate numbers and through appropriate mechanisms to provide viable and sustainable benefits to the selected recipients.

Key actions

- Ensure the age, sex, number, breed and animal types provided are appropriate and adapted to local agro-ecological and climatic conditions *(see Guidance note 1)*.

- Ensure that selected animals are healthy and productive *(see Guidance note 2)*.

- Base selection of recipients on locally agreed criteria *(see Guidance note 3)*.

Guidance notes

1. Selection of animals

It is important that the age, sex, number, breed and types of the animals that are provided meet the livelihood needs of the planned recipients. This should take into account the different priorities of men and women. Typically, young female livestock of breeding age are prioritised, together with single breeding males. By doing this, recipients are given every opportunity to rebuild their livestock assets. However, where the focus is on the provision of animals for transport or ploughing, then male animals may be selected.

In pastoral areas, it is important that interventions provide adequate numbers of core breeding animals to establish a viable flock or herd over time *(see Appendix 9.3: Discussion on minimum viable herd size*; see also *Process case study: Community contributions to herd replacement in Ethiopia)*. Ideally, animal numbers should be tailored to individual household size. However, experience suggests that it is more common for agencies to provide a standard number of animals to each household irrespective of household size. In this way, all households are treated equally, and the costs are similar.

Local breeds with which livestock keepers are familiar are likely, over time, to outperform imported and exotic animals, which do not thrive under poor nutrition and management conditions. (See *FAO, 2016* pp. 146–151 on specifications of animals and the inspection criteria). Local purchase supports local livestock keepers and markets. Local purchase also avoids potentially complex logistical, health-related, environmental, welfare and financial issues associated with the movement of animals from distant areas and neighbouring countries.

An emergency that occurs at scale and severely reduces livestock numbers in an area may make it impossible to source adequate numbers of livestock locally. In such cases, agencies may need to purchase animals from greater distances. Here, they should take care to ensure all purchased animals are adapted to local agro-ecological conditions, including climatic and seasonal disease patterns. Such animals are more likely to be able to thrive and be productive.

In conflict situations, or areas of insecurity where looting is common, agencies should ensure that no looted animals are included in the distribution. If there is any doubt, the distribution should be delayed.

2. Veterinary inspection

Animals that are selected for the intervention should all be healthy and productive or, where immature animals are provided, likely to be productive in the future. At the time of purchase, a trained veterinarian or veterinary paraprofessional should inspect animals. Inspection typically takes place at the point of purchase, such as a market or at a designated facility such as a quarantine facility or holding ground. The inspector may be a local private practitioner contracted by the response agency, or a government official. The inspection should highlight any key disease issues, and any unsuitable animals should be excluded. Selected animals may be marked for identification purposes using ear tags, ear notching, or branding. The identification, handling, and marking of animals should meet high welfare standards. See *Chapter 1: Introduction to LEGS* on animal welfare, and *FAO* (2016, pp. 151–152) on livestock identification.

3. Recipient selection

Recipient selection should assess the necessary skills, capacities, experience and access to productive resources. Those without previous experience of livestock keeping should demonstrate a real interest, have access to productive resources, and be committed to structured learning. Where there are questions about a potential recipient's interest and commitment, then the household should be deselected. Such decision-making processes are generally only effective when fully supported by customary and other local community leaders. See also *FAO* (2016, p. 148) on roles and responsibilities of restocking committees.

Technical standards
Provision of livestock

Standard 3: Delivery systems

The delivery of livestock to recipients is efficient, cost-effective and supports the earliest possible return to livestock production and productivity.

Key actions

- Ensure that procurement meets agreed criteria and all legal procurement procedures are observed *(see Guidance note 1)*.
- Make full use of cash, vouchers and livestock fairs *(see Guidance note 2)*.
- Only provide livestock under a credit system in special circumstances. In all other cases, provide livestock as a gift *(see Guidance note 3)*.
- Where possible, use customary livestock redistribution mechanisms as a guide *(see Guidance note 4)*.
- Ensure the timing of animal provision is seasonally appropriate and that animals can continue to grow and be productive *(see Guidance note 5)*.
- Plan transport in advance to minimise risk of losses in transit, based on conditions that ensure the welfare of the stock *(see Guidance note 6)*.

Guidance notes

1. Procurement procedures

It is important that agencies involved in the provision of livestock observe all regulations concerning the purchase and distribution of livestock, including taxes, quarantine, cross-border issues, etc. Such regulations may impact the efficiency of the intervention. For example, taxes increase costs, and quarantine requirements (where stringent disease control and certification measures have been put in place) both delay distribution and increase costs. See *FAO* (2016, pp. 141–144) on animal healthcare, quality control and use of holding grounds.

2. Use of cash, vouchers and livestock fairs

Where possible, involve the recipients themselves in the selection of livestock, through cash, voucher and livestock fair mechanisms. Whether the livestock are selected by the recipients themselves, or by representatives (local experts, traders or elders), the use of cash, vouchers and livestock fairs helps ensure that only the right type of animals are selected for the intervention. In recent times, cash, voucher and livestock fair-based distributions have increasingly been replacing direct distribution.

Ahead of such distribution mechanisms, it is necessary for implementing agencies to assess local market prices and to set livestock values accordingly. This will ensure that total purchases can be supported by available intervention funds.

3. Credit systems that do not jeopardise productivity

During the design stage, the decision should be made as to whether the intervention will be based on credit or on gift distribution. If credit, it should also be decided what form the repayment should take. This should be done in close consultation with the recipients and based on full understanding and commitment from all participating households. Where livestock are provided under a credit system, the loan may either be repaid in the form of offspring of the livestock, or in cash. Cash repayment requires a level of community integration into a market economy. So, in many cases, repayment in the form of stock will be more appropriate, and more likely to build on customary systems.

The repayment arrangement (type and condition of animal, timing of repayment, etc.) must be planned carefully. The aim here is to ensure repayment does not negatively affect the quality of livelihood support received from the initial livestock provision. For example, if the animals provided are not productive, the repayment can burden the recipient with a debt. Selection of secondary recipients (to receive cash or livestock offspring from primary recipients) should take place at the time that primary recipients are identified, and repayment should be carefully monitored. See *FAO* (2016, p. 136) on the advantages and disadvantages of providing animals as gifts or loans.

4. Customary redistribution systems

Livestock are redistributed in pastoral communities through marriage, between family members and friends, and as social support – gifts and loans – to households in need of additional productive animals. For example, a milk camel may be loaned to a family with no milking camel for a period of months to help provide that household with adequate milk. Wherever possible, provision of livestock interventions should support, strengthen and be aligned with such customary redistribution mechanisms. For example, the selection of appropriate recipients can be done with customary leaders, provided issues of inclusion of women and marginal groups are also agreed. In such ways, customary redistribution institutions and mechanisms can continue to operate after the intervention is complete. Where possible and appropriate, the selection of recipients should be integrated into customary livestock redistribution systems and practices.

5. Timing of distribution

Whichever mechanism is used, local knowledge can be used to ensure that livestock are provided at the start of the season when naturally occurring feed and water is most widely available. In this way, animal growth and productivity are maximised and any negative environmental impacts minimised. According to seasonal disease calendars, animals should be vaccinated against locally occurring seasonal diseases that might affect production and productivity.

6. Transport planning

Transport for livestock, including itinerary, duration, likely weather conditions, distances, opening hours of customs, staging points, and sufficient rest periods to provide water and feed (and for milking if necessary), needs to be planned well in advance. The equipment and supplies needed to feed, water and milk the stock as necessary should be planned for as well. The conditions and length of the journey should ensure the welfare of the livestock as described under the five domains *(see Chapter 1: Introduction to LEGS)*. It is important for those involved in transport to avoid overloading (and the resultant risk of suffocation) and ensure sufficient space for animals to stand and lie in their normal position. Animals should at the same time be packed closely (as appropriate for the species) to avoid falling. The vehicle should be disinfected before and after loading, and be properly ventilated. The delivery site should also be properly prepared with sufficient water, feed, fencing and shelter.

Technical standards
Provision of livestock

Standard 4: Additional support to recipients

Additional support is provided to recipients, in the form of veterinary care, training and other livelihood support, to help ensure a positive and sustainable impact on livelihoods.

Key actions

- Provide preventive veterinary care for the livestock prior to distribution *(see Guidance note 1)*.
- Establish a system for the ongoing provision of veterinary care *(see Guidance note 2)*.
- Provide training and capacity-building support to recipients based on an analysis of skills and knowledge of animal husbandry *(see Guidance note 3)*.
- Ensure that training and capacity building includes preparedness for future shocks and emergencies *(see Guidance note 4)*.
- Identify and meet food security needs according to the Sphere 'Minimum standards in food security and nutrition' to prevent early offtake of livestock *(see Guidance note 5)*.
- Identify and meet shelter and non-food needs according to the Sphere standards on 'Shelter and settlement' *(see Guidance note 6)*.
- Withdraw food security support only when herd size and/or the emergence of other economic activities enable independence from such support *(see Guidance note 7)*.

Guidance notes

1. Preventive veterinary care

Prior to distribution, animals should be vaccinated against locally important infectious diseases, dewormed and receive other preventive animal healthcare to remain healthy and productive. Typically, this service is provided as a single, free-of-charge input *(See Process case study: Livestock distribution to at-risk households following an earthquake in Pakistan and Impact case study: Post-earthquake livestock distribution in Iran.)* There may also be government requirements on vaccines, drenching, deworming, etc. that will need to be checked.

2. Long-term veterinary care

Recipient households should have continued access to preventive and curative animal healthcare services until livestock are contributing fully to livelihoods. This may take some months or even longer. Delivery of this service may be through one of a range of different animal health service providers. These include private community-based animal health workers, private veterinarians, private veterinary livestock medicine outlets, or public government veterinary services. This long-term care system may also provide an opportunity for implementing agencies to collect monitoring and evaluation data. The provision of animal healthcare services and associated training should meet the standards and guidelines set out in *Chapter 6: Veterinary support*.

3. Training and capacity strengthening

Training in animal husbandry may not be necessary for livestock replacement activities where recipients already have considerable knowledge and experience in livestock management. However, in other cases, households are new to livestock keeping; knowledge and skills have been lost; or new technologies have become available. Here, training is imperative to help increase survival and productivity levels, and contribute to improved outcomes. Training and/or the provision of information on livestock sector markets and value chains may also be useful to secure more viable livestock-based livelihoods in the longer term.

4. Preparedness for future emergencies

In communities with little livestock management experience, it is helpful for agencies to invest in building emergency preparedness knowledge and skills. This will minimise the risk of future losses. Training may include improved forage and feed production, feed storage, animal healthcare, livestock marketing (including livestock offtake), shelter improvements, and protection and maintenance of water sources *(see Chapters 4–8)*.

5. Food security support

Households that receive livestock may continue to experience high levels of food shortages until livestock numbers and production, or crop production, return to pre-emergency levels. Until this happens, recipient households may need to consume or sell distributed animals to meet acute household needs. For this reason, food security needs of recipient households should be assessed, and additional support provided until the livestock become fully productive. The Sphere Handbook provides 'Minimum standards for food security and nutrition'. Cash and voucher assistance may be appropriate for providing this support.

6. Shelter and non-food support

Families receiving livestock may also require shelter, basic household utensils, bedding, water containers, and livestock-related equipment such as carts, harnesses and ploughs. Without such support, recipients may either be forced to sell livestock to meet other needs, or be unable to make full use of distributed livestock. The use of CVA to meet these needs can be appropriate, as this type of assistance allows individuals, households and communities to set their own priorities.

7. Withdrawal of food security support

Agencies can withdraw food security assistance once the herd size grows, or once other livelihood activities provide enough support to recipient households so they can avoid unnecessary livestock offtake. A well-designed participatory monitoring system can include measures of herd growth and other livelihoods-based indicators to determine the best time for them to withdraw this support.

Technical standards
Provision of livestock

Appendices

Appendix 9.1: Assessment checklist for provision of livestock
Options and implications
- What role did livestock play in livelihoods before the emergency?
  - Main livelihood asset?
  - Provision of supplementary food?
  - Income generation?
  - Transport or draught power?
- Which species and breeds were kept? By whom? And for what purposes?
- Which species and breeds have been lost and need replacement?
- If livestock did not already form part of livelihood strategies:
  - Is there potential for the introduction of livestock to meet supplementary food or income-generation needs?
  - Which species and breeds would be most appropriate for distribution?
- Have alternative, more cost-effective options than livestock provision been considered?
- What customary mechanisms exist for redistributing livestock?
- What numbers of livestock would constitute the minimum viable herd per household in the local context?
- What are the implications of distributing these minimum numbers of livestock in the area?
  - Is there sufficient pasture or feed?
  - Is there sufficient water?
  - Is there adequate shelter, or can this be constructed?
  - Will the livestock be secure, or will the activity increase the risk to livestock keepers and/or the animals themselves?
  - Are animal health services available?

Recipients

- What social, physical and natural livelihood assets do potential recipient households have to enable them to manage livestock successfully in the future?
- Can training in livestock management be provided if necessary?
- What roles do women and men play in livestock management and care? Do women want to take on new roles and responsibilities for livestock/production? What are the labour implications of livestock provision?
- What are the particular needs of at-risk groups in relation to livestock management and access to livestock products?
- Are there sufficient resources to provide livestock-related support to recipient households (for example, veterinary care, feed, shelter) as required?
- Are there sufficient resources to provide non-livestock support to recipient households as required (for example, food or other livelihood support while herds rebuild)?

Procurement

- What are the implications of the purchase of significant numbers of livestock on local markets?
- Are livestock available for purchase in sufficient numbers within transporting distance of recipient communities?
- Is transport available, and can stock be transported safely without risk to their welfare?
- What are the risks of disease from importing stock from another area?

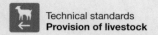

**Technical standards
Provision of livestock**

Appendix 9.2: Examples of monitoring and evaluation indicators for the provision of livestock

	Process indicators (measure things happening)	Impact indicators (measure the result of things happening)
Designing the intervention	Number of meetings with community representatives and other stakeholders, including private sector suppliers where relevant Meetings held include the views of the various gender-age groups	Meeting reports with analysis of options for livestock provision Action plan including and differentiating the priorities of various gender-age groups: • roles and responsibilities of different actors • community process and criteria for selecting recipient households • community preferences for livestock species and type • procurement, transportation, and distribution plan, with recipient households' involvement • veterinary inspection and preventive care

continued over

Technical standards
Provision of livestock

	Process indicators (measure things happening)	Impact indicators (measure the result of things happening)
Replacing lost livestock assets: provision of core breeding animals for pastoralists and agro-pastoralists	Number of livestock provided per household by livestock type	Mortality in livestock provided versus mortality in pre-existing livestock
	Type and value of additional support to each household (e.g. food aid, utensils, etc.)	Number of offspring from livestock provided, and uses of offspring (e.g. sales and use of income)
	The participation (voice) of women in the implementation of the intervention	Human nutrition – consumption of milk by children in households receiving livestock
		Herd growth and levels of reliance on external assistance over time
		Changes in women's and children's roles and responsibilities in livestock management (comparison of before and after livestock provision)
		Changes in the workload of women (e.g. after provision of working animals)
		Note: compare impacts between livestock recipients and non-recipients

continued over

Technical standards
Provision of livestock

	Process indicators (measure things happening)	Impact indicators (measure the result of things happening)
Replacing lost livestock assets: smallholder farmers and other income generation	Number of livestock provided per household by livestock type	Mortality in livestock provided versus mortality in pre-existing livestock
	Type and value of additional support to each household (e.g. food aid, utensils, etc.)	Number of offspring from livestock provided, and uses of offspring (e.g. sales and use of income)
	Training, where appropriate, on livestock production and management	Human nutrition – consumption of milk by children in households receiving livestock
	The participation (voice) of women in the implementation of the intervention	Changes in women's and children's roles and responsibilities in livestock management (a comparison of before and after livestock provision)
		Changes in the workload of women (e.g. after provision of working animals)
		Note: compare impacts between livestock recipients and non-recipients

continued over

Technical standards
Provision of livestock

	Process indicators (measure things happening)	Impact indicators (measure the result of things happening)
Building livestock assets	Number of livestock provided per household by livestock type	Mortality in livestock provided versus mortality in pre-existing livestock
	Training on livestock production, management and marketing	Number of offspring from livestock provided, and uses of offspring (e.g. sales and use of income)
	The participation (voice) of women in the implementation of the intervention	Human nutrition – consumption of milk by children in households receiving livestock
		Changes in women's and children's roles and responsibilities in livestock management (a comparison of before and after livestock provision)
		Changes in the workload of women (e.g. after provision of working animals)
		Note: compare impacts between livestock recipients and non-recipients

Technical standards
Provision of livestock

Appendix 9.3: Discussion on minimum viable herd size

Herd replacement interventions in pastoralist areas often use the concept of 'minimum viable herd size'. This is to determine the minimum number and types of animals required to allow pastoralists to maintain a pastoralism-based livelihood. It might be convenient for standards and guidelines such as LEGS to indicate a specific number and type of animals to be provided. Yet this differs significantly between pastoralist groups, and there are no standard numbers of livestock that should be given. Similarly, in mixed farming communities, it is difficult to determine a global figure for livestock provision.

If pastoralists completely depend on livestock, then a minimum herd size would need to be around 2.5 to 3 Tropical Livestock Units (TLUs)/adult equivalent, according to the role played by livestock. (1 camel = 1TLU; 1 head of cattle = 0.7 TLUs; 1 sheep or goat = 0.1 TLUs). Talks with local pastoralists confirm this. If this level of transfer is not achieved, then implementing agencies would need to support the household by cash/food transfers. Otherwise, they may be forced to a) sell animals periodically; or b) send members of the family to live with relatives; or c) require remittances on a regular basis.

The best way to determine how many and which types of livestock to provide is by undertaking participatory analysis and discussion with the communities concerned. This process may include a description of the benefits and problems of different livestock species and breeds for the different wealth, gender and age groups within the community. It would also include an analysis of any customary restocking systems. A further consideration is that although a 'minimum herd size' may be defined with communities through such analysis/discussion, many agencies are faced with limited budgets for providing livestock. So, the more animals provided per household, the fewer the total number of households that will benefit from the initiative.

Save the Children UK implemented a restocking project between 2002 and 2003 for 500 internally displaced families in eastern Ethiopia as a post-drought response. It provided each pastoral household with 30 breeding sheep or goats. The project was implemented in collaboration with the Ethiopian government's Disaster Preparedness and Prevention Committee and the Somali region livestock bureau. The total budget was around US$244,500 – equivalent to $489 per household. This budget excluded the cost of food aid and household items, which were provided by other agencies such as the Christian Relief and Development Association and UNICEF.

An evaluation concluded that the project had provided substantial benefits through the restocking process. However, it observed that the package should have included at least 50 sheep and goats per household in order for the families to have a viable source of livelihood. This would have increased the project budget by 41 per cent if 500 households were still to be targeted. Alternatively, the original budget could have covered 300 households with 50 animals each. The evaluation indicated that a budget of around $690 per household was needed to restock the target communities in a viable way (Wekesa, 2005). This example illustrates the challenge faced by aid agencies when deciding how many households to restock and how many animals to provide. It also shows the importance of determining the appropriate definition of 'minimum viable herd' in each specific context.

References and further reading

General

FAO (2016) *Livestock-related interventions during emergencies – The how-to-do-it manual*. Edited by Philippe Ankers, Suzan Bishop, Simon Mack and Klaas Dietze. FAO Animal Production and Health Manual No. 18. Rome, https://www.fao.org/3/i5904e/i5904e.pdf (See Chapter 9)

Heffernan, C., Misturelli, F. and Nielsen, L. (2004) *Restocking Pastoralists: A Manual of Best Practice and Decision Support Tools*, Practical Action Publishing, Rugby, https://practicalactionpublishing.com/book/1789/restocking-pastoralists

Knight-Jones, T. (2012) *Restocking in the former Yugoslavia: post-war restocking projects in Bosnia-Herzegovina and Kosovo*, World Society for the Protection of Animals (WSPA), London, https://www.livestock-emergency.net/wp-content/uploads/2020/04/Balkans_Restocking_Final2.pdf

Toulmin, C. (1995) 'Tracking through drought: options for destocking and restocking', in I. Scoones (ed.), *Living with Uncertainty: New Directions in Pastoral Development in Africa*, pp. 95–115, Intermediate Technology Publications, London.

Traditional livestock redistribution

Lotira, R. (2004) *Rebuilding Herds by Reinforcing Gargar/Irb among the Somali Pastoralists of Kenya: Evaluation of Experimental Restocking Program in Wajir and Mandera Districts of Kenya*, African Union/Interafrican Bureau for Animal Resources, Nairobi, https://fic.tufts.edu/wp-content/uploads/Lotira-Restocking-evaluation.pdf

Cash transfers for herd replacement

O'Donnell, M. (2007) *Cash-based emergency livelihood recovery programme, Isiolo District, Kenya*, project evaluation draft report, Save the Children, Nairobi, https://www.livestock-emergency.net/wp-content/uploads/2020/05/ODonnell-2007.pdf

FAO (2011) *The use of cash transfers in livestock emergencies and their incorporation into Livestock Emergency Guidelines and Standards (LEGS)*, Animal Production and Health Working Paper No. 1, FAO, Rome, https://www.fao.org/3/i2256e/i2256e00.pdf

Impact of herd replacement

Budisatria, I.G.S. and Udo, H.M.J. (2013) Goat-based aid programme in Central Java: an effective intervention for the poor and vulnerable? *Small Ruminant Research* 109: 76-83, https://www.livestock-emergency.net/wp-content/uploads/2020/04/Budisatria-and-Udo-2013.pdf

Croucher, M., Karanja, V., Wako, R., Dokata, A. and Dima, J. (2006) *Initial impact assessment of the livelihoods programme in Merti and Sericho*, Save the Children Canada, Nairobi.

Heffernan, C. and Rushton, J. (2000) 'Restocking: A critical evaluation', *Nomadic Peoples* 4(1):110–124, https://www.researchgate.net/publication/237632191_Restocking_A_Critical_Evaluation

Leguene, P. (2004) *Evaluation report: restoration of the livelihood and longer-term food security for the earthquake-affected farmers and agricultural labourers in Bam, south-sast Iran*, project implemented by ACF-Spain &, ACF-UK London.

Sadler, K., Mitchard, E., Abdulahi, A., Shiferaw, Y., Bekele, G. and Catley, A. (2012) *Milk Matters: The Impact of Dry Season Livestock Support on Milk Supply and Child Nutrition in Somali Region, Ethiopia*, Feinstein International Center, Tufts University and Save the Children, Addis Ababa, http://fic.tufts.edu/publication-item/milk-matters/

Save the Children International (2013) *Drought Early Warning and FSL Needs Assessment in Hiran and Puntland*; *Livestock Baseline for Hiran, DFID Project*; *Evaluation of Livelihoods/Resilience Activities, Hiran*; and *Livestock and Cash Grants Project Baseline for Hiran*, Save the Children International, Nairobi.

Wekesa, M. (2005) *Terminal Evaluation of the Restocking/Rehabilitation Programme for the Internally Displaced Persons in Fik Zone of the Somali Region of Ethiopia*, Save the Children UK, Addis Ababa and Acacia Consultants, Nairobi.

Animal diseases and restocking

Knight-Jones, T. (2012) *Restocking and Animal Health: A Review of Livestock Disease and Mortality in Post-Disaster and Development Restocking Programmes*, World Society for the Protection of Animals (WSPA), London, https://www.researchgate.net/publication/261511178_Restocking_and_animal_health_A_review_of_livestock_disease_and_mortality_in_post-disaster_and_development_restocking_programmes_WSPA

Restocking with exotics

Heffernan. C (2009) 'Biodiversity versus emergencies: the impact of restocking on animal genetic resources after disaster', *Disasters* 2009, 33(2): 239-252 https://doi.org/10.1111/j.1467-7717.2008.01072.x

Donkeys and the provision of livestock

Catley, A. and Blakeway, S. (2004) 'Donkeys and the provision of livestock to returnees: lessons from Eritrea', in P. Starkey and D. Fielding (eds), *Donkeys, People and Development: A Resource Book of the Animal Traction Network for Eastern and Southern Africa* (ATNESA), pp. 86–92, Technical Centre for Agricultural and Rural Cooperation, Wageningen, The Netherlands, http://www.atnesa.org/donkeys/donkeys-catley-returnees-ER.pdf

See also case studies for provision of livestock in emergencies at: https://www.livestock-emergency.net/resources/case-studies/

Annexes

- 364 Annex A: Glossary
- 374 Annex B: Abbreviations and acronyms
- 375 Annex C: Combined timing table
- 376 Annex D: Acknowledgements and contributors
- 379 Index

Annex A: Glossary

Acaricide
A chemical used to kill ticks, for example, in a spray, pour-on, or as a dip solution.

Accountability
'Accountability towards affected people is the process of using power responsibly. (It is) taking account of, and being held accountable by, different stakeholders, primarily those who are affected by the exercise of such power.' (CHS Alliance)

Affected communities
Communities who are affected, either directly or indirectly, by a hazardous event. Directly affected are those who have suffered injury, illness or other health effects; who were evacuated, displaced or relocated; or who have suffered direct damage to their livelihoods or their economic, physical, social, cultural or environmental assets. (Based on UNDRR)

Agencies
LEGS uses this term to refer to all who respond to an emergency. It includes government agencies; local, national and international organisations; and community groups.

Agro-pastoralists
Agro-pastoralists integrate seasonal crop production with *pastoralist* livestock rearing.

Alarm phase
The second phase of a *slow-onset emergency*.

Alert phase
The first phase of a *slow-onset emergency*.

Anthelmintic
A veterinary drug used to kill parasitic worms.

At risk, at-risk
Likely to be in danger of serious harm (loss of life, livelihood, assets, or services) because of a combination of external threats (natural hazards, gender-based violence, etc.) and individual vulnerabilities (poverty, membership of a marginalised group etc.). (Drawn from Sphere)

Benefit-cost analysis

'A technique used to compare the costs of an intervention with its benefits, using a common metric (usually a monetary unit) to calculate its net cost or benefit.' (CALP Network)

Cash and voucher assistance (CVA)

'Cash and voucher assistance (CVA) refers to the direct provision of *cash transfers* and/or *vouchers* for goods or services to individuals, households or group/community recipients. In the context of humanitarian response, CVA excludes payments to governments or other state actors, *remittances*, service provider stipends, microfinance and other forms of savings and loans.' (CALP Network)

Cash for work (CFW)

'Cash payments provided on the condition of undertaking designated work. This is generally paid according to time worked (e.g. number of days, daily rate), but may also be quantified in terms of outputs (e.g. number of items produced, cubic metres dug). CFW interventions are usually in public or community work programmes but can also include home-based and other forms of work.' (CALP Network)

Chronic/recurring emergency

A *slow-onset emergency* in which the phases (*alert*, *alarm*, *emergency*, *recovery*) keep repeating themselves without returning to 'normal'.

Cold chain

A system whereby veterinary or human medicines are kept at the required low temperature during storage and transportation using refrigerators and mobile cold boxes.

Complex emergency

'A humanitarian crisis in a country or region in which authority has totally or substantially broken down due to multiple causes and where people's lives, wellbeing and dignity are affected. The crisis may have been caused by human activity (i.e., conflict or civil unrest) and/or by natural factors (e.g., drought, flood, hurricanes).' (Sphere)

Conditions/conditional/conditionality

'Conditionality refers to prerequisite activities or obligations that a recipient must fulfil to receive assistance. Conditions can be used with any kind of transfer (cash, vouchers, in-kind, service delivery) depending on the intervention design and objectives.' (CALP Network)

Core breeding animals

Those animals within the herd/flock that livestock owners identify as essential for rebuilding their stock. They are likely to include the most fertile and productive animals, as well as those with desirable traits such as a good temperament. The core breeding herd will be age-structured to include animals of different maturity levels.

Crisis modifiers

Financial mechanisms built into multi-year grants that are designed to be released when agreed early warning emergency triggers are met, enabling rapid response.

Disaster

See 'Emergency'.

Disaster risk management (DRM)

The actions needed to achieve the objective of reducing risk. The DRM cycle incorporates risk identification, mitigation, prevention and preparedness, as well as the response, recovery and reconstruction activities that are undertaken following the disaster [emergency].

Disaster risk reduction (DRR)

'Reducing the risk of disaster [emergency] through systematic efforts to analyse and manage causal factors: […] reducing exposure to hazards, lessening the vulnerability of people and property, wise management of land and the environment, and improving preparedness for adverse events.' (Sphere)

Drought cycle management model

A model for drought response that divides drought into four phases *(alert, alarm, emergency* and *recovery)*. These terms are used by LEGS as the four phases of a *slow-onset emergency*.

Dzud

Severe climatic conditions in Mongolia (summer drought followed by a severe winter) in which large numbers of livestock die due to lack of grazing.

Early action
'Early action, also known as anticipatory action or forecast-based action, means taking steps to protect people before a disaster [emergency] strikes, based on early warning or forecasts. To be effective, it must involve meaningful engagement with at-risk communities.' (IFRC)

Emergency
'A calamitous event that seriously disrupts the functioning of a community or society and causes human, material, and economic or environmental losses that exceed the community's or society's ability to cope using its own resources.' (IFRC)

Emergency phase
The third phase of a *slow-onset emergency*.

Early recovery phase
The second phase of a *rapid-onset emergency*.

Evaluation
A periodic/planned assessment of an intervention to review its effectiveness, relevance, efficiency, impact and sustainability. Evaluations may focus on whether the intervention's objectives have been achieved and the factors that influenced the results, or may assess its livelihood impacts using impact indicators.

Gender
'The roles, responsibilities and identities of women and men and how these are valued in society. These vary in different cultures and change over time. Gender identities define how society expects women and men to think and act. Gender roles, responsibilities and identities can be changed because they are socially learned.' (Sphere)

Guidance notes
To be read in conjunction with the *key actions*. They explain particular issues and how to address any practical difficulties when applying the LEGS *Standards*.

Hazard
'A process, phenomenon or human activity that may cause loss of life, injury or other health impacts, property damage, social and economic disruption or environmental degradation.' (UNDRR)

Annexes

Humanitarian crisis
A serious disruption of the functioning of a community or a society involving widespread human, material, economic or environmental losses and impacts. These losses/impacts exceed the ability of the affected community or society to cope using its own resources, and therefore require urgent action. Related terms are *disaster* and *emergency*.

Immediate aftermath phase
The first phase of a *rapid-onset emergency*: the period just after the emergency has struck, when the impact is greatest.

Impact indicator
Point of reference for measuring the result of actions taken in terms of their effect on recipients.

Inclusion
'A rights-based approach to community programming, aiming to ensure all people who may be at risk of being excluded have equal access to basic services and a voice in the development and implementation of those services. At the same time, it requires that organisations make dedicated efforts to address and remove barriers to access services.' (Sphere)

The LEGS Handbook highlights that those at risk of exclusion may include children (in particular separated, unaccompanied or orphaned children), women, older people, people with disabilities, or groups marginalised because of religion, ethnic group, caste or gender identity.

Indicators
Measurements (either qualitative or quantitative) of the progress of an intervention. LEGS divides them into *process indicators* and *impact indicators*.

Internally displaced person
Someone who is forced to leave their home but who remains within their country's borders.

Intervention
The package of activities designed to contribute to an outcome. In LEGS, the interventions cover the specific technical areas selected as the best response to an emergency situation (feed, water, veterinary support, shelter, livestock offtake and provision of livestock). Each intervention is broken down into different *options*.

Key actions
Practical steps or actions for achieving the LEGS *Standards*, not all of which may be relevant to all situations.

Learning
Gathering, analysing and disseminating information from monitoring, evaluation and accountability processes, and ensuring this is proactively applied to current and future planning.

Livelihoods
The capabilities, assets and activities required to make a living.

Livelihood assets
The resources, equipment, skills, strengths and relationships that are used by individuals and households to pursue their livelihoods. They are categorised as social, human, natural, financial and physical.

Livestock offtake
Animals sold to traders or otherwise removed from the herd/flock.

Livestock settlement
The term used to refer to the wider context of livestock shelters, including land rights, environmental implications of shelter provision, and access to feed and water.

Livestock shelters
The physical structures that, in some environments, animals need to survive. Shelters can be either temporary or longer-lasting.

Lower- and middle-income countries
The World Bank classifies economies for analytical purposes into four income groups: low, lower-middle, upper-middle and high-income. Countries are classified each year based on gross national income per capita in US dollars (see World Bank). Based on this, LEGS uses the term 'lower- and middle-income countries'.

Monitoring
A continuous/systematic process for keeping interventions on track, using the intervention's process indicators to provide timely information for decision-making.

One Health
'An integrated, unifying approach that aims to sustainably balance and optimise the health of people, animals and ecosystems.' (WHO, FAO, UNEP and WOAH)

Options
Each LEGS technical intervention is divided into different options, which present different ways of delivering a technical response (e.g. water trucking versus borehole construction).

Participatory methods and approaches
Techniques which, based on a defined methodology and systematic learning process, focus on cumulative learning by all participants, draw on multiple perspectives, and involve group learning processes. (Drawn from IIED)

See *Appendix 3.1* for more details.

Participatory epidemiology
The use of participatory approaches to work with communities to study specific disease problems and solutions.

Participatory targeting
The use of participatory approaches to help ensure that interventions focus on those who, as the community has agreed, need them most. LEGS commitment to the Core Humanitarian Standard requires that affected communities must participate in decisions about emergency responses.

Pastoralists
Pastoralists are largely dependent for their food and income on extensive livestock rearing.

Peri-urban
The area surrounding a city or town that forms the interface between urban and rural land usage and is dependent on both – for example, it may have a market selling rural products to urban consumers.

Phytosanitary
Relating to plant health and biosecurity; the WTO 'Agreement on the Application of Sanitary and Phytosanitary Measures' sets out the basic rules for food safety and animal and plant health standards.

Process indicator

Point of reference for measuring the implementation of an intervention over a defined timeframe. Process indicators are usually quantitative (also called progress indicators).

Prophylactic drugs

A medicine or treatment designed and used to prevent disease.

Protracted emergency

'Protracted emergencies are situations where a significant part of the population is acutely vulnerable and dependent on humanitarian assistance over a prolonged period of time. In many cases, this period becomes so long that the emergency has become the normal situation.' (Danish Refugee Council)

Purposive samples

The selection of a 'typically' representative group based on particular characteristics (for example, livestock owners affected by drought; women livestock keepers; inhabitants of a flood- affected village).

Rapid-onset emergency

An emergency such as an earthquake, flood, or tsunami that occurs very suddenly and sometimes without warning. LEGS divides rapid-onset emergencies into three phases – *immediate aftermath*, *early recovery* and *recovery*.

Recovery phase

The third phase of a *rapid-onset emergency* or the fourth phase of a *slow-onset emergency*.

Resilience

'The ability of an individual, community, society or country to anticipate, withstand and recover from adversity – be it a natural hazard or other emergency. Resilience depends on the diversity of livelihoods, coping mechanisms and life skills such as problem-solving, the ability to seek support, motivation, optimism, faith, perseverance and resourcefulness.' (Sphere)

Response

'The provision of services and assistance during or immediately after a disaster [emergency] in order to save lives, reduce health impacts, ensure public safety, maintain human dignity and meet the basic subsistence needs of the people affected.' (Sphere)

Restrictions

'Restriction refers to limits on the use of assistance by recipients. Restrictions apply to the range of goods and services that the assistance can be used to purchase, and the places where it can be used. Vouchers are restricted transfers by default, since they are inherently limited in where, when and how they can be used.' (CALP Network)

SMART objectives

Response plan objectives should be specific, measurable, achievable, relevant, and time-bound (SMART), so that they quantify the intended livelihood impacts within a specific time frame.

Slow-onset emergency

An emergency, such as a drought or extreme cold season, whose effects are felt gradually. LEGS divides these into four key phases – *alert*, *alarm*, *emergency* and *recovery*.

Standards

Qualitative statements of the minimum to be achieved in any emergency in any context.

Theory of change

A description or illustration that explains how an intervention is expected to achieve its intended final outcome. It draws on causal analysis and underlying assumptions to show how a series of results will contribute to the proposed change.

Veterinary paraprofessionals

Any veterinary worker who works under the supervision of a veterinarian, performs designated professional tasks and receives training accordingly. Some countries categorise community-based animal health workers (CAHWs) as veterinary paraprofessionals, and some do not. The term 'veterinary paraprofessionals' in LEGS includes CAHWs.

Zoonosis

A disease that can be transmitted from animals to humans. Related terms: zoonotic disease, zoonoses (pl)

Annexes

Sources

CALP Network – https://www.calpnetwork.org/resources/glossary-of-terms/

CHS Alliance – https://www.chsalliance.org/accountability-to-affected-people/

Danish Refugee Council – https://emergency.drc.ngo/crisis/emergency-typologies/

IFRC – https://www.ifrc.org/early-warning-early-action

IIED – https://www.iied.org/sites/default/files/pdfs/migrate/6021IIED.pdf

Sphere – https://spherestandards.org/wp-content/uploads/Sphere-Glossary-2018.pdf

UNDESA – https://www.un.org/esa/socdev/rwss/2016/full-report.pdf

UNDRR – https://www.undrr.org/terminology

WHO, FAO, UNEP and WOAH – https://www.who.int/news/item/01-12-2021-tripartite-and-unep-support-ohhlep-s-definition-of-one-health

World Bank – https://datatopics.worldbank.org/world-development-indicators/the-world-by-income-and-region.html

Annex B: Abbreviations and acronyms

AMR	antimicrobial resistance
CAHW	community-based animal health worker
CCCM	Camp Coordination and Camp Management
CFW	cash for work
CHS	Core Humanitarian Standard on Quality and Accountability
CPMS	Child Protection Minimum Standards
CVA	cash and voucher assistance
DRM	disaster risk management
DRR	disaster risk reduction
FAO	Food and Agriculture Organization of the United Nations
GIS	geographic information systems
HIS	Humanitarian Inclusion Standards
HSP	Humanitarian Standards Partnership
HLP	housing, land and property rights
ID	identity document
IDP	internally displaced person/people
INEE	Inter-Agency Network for Education in Emergencies
LEGS	Livestock Emergency Guidelines and Standards
MEAL	monitoring, evaluation, accountability and learning
MERS	Minimum Economic Recovery Standards
MISMA	Minimum Standards for Market Analysis
MNB	multi-nutrient block
OECD	Organisation for Economic Co-operation and Development
PRIM	LEGS Participatory Response Identification Matrix
SEADS	Standards for supporting crop-related livelihoods in emergencies
SMART	specific, measurable, achievable, relevant and time-bound (indicators)
ToC	theory of change
WASH	water supply, sanitation and hygiene
WHO	World Health Organization
WOAH	World Organisation for Animal Health (formerly OIE)

Annex C: Combined timing table

Figure A.1: Combined timing table

Technical interventions and options	Rapid-onset emergency			Slow-onset emergency			
	Immediately after	Early recovery	Recovery	Alert	Alarm	Emergency	Recovery
Feed							
Home-based	✓	✓	—	—	✓	✓	—
Feed camp	✓	✓	—	—	✓	✓	—
Water							
Water points: changing management	✓	✓	✓	✓	✓	✓	✓
Water points: rehabilitating	✓	✓	✓	✓	✓	✓	✓
Water points: establishing	—	—	✓	✓	✓	—	✓
Water trucking	✓	—	—	—	—	✓	—
Veterinary support							
Clinical veterinary services	✓	✓	✓	✓	✓	✓	✓
Public sector veterinary functions	—	—	✓	—	—	—	✓
Shelter							
Temporary shelter interventions	✓	—	—	—	✓	✓	—
Longer-lasting shelter interventions	✓	✓	✓	✓	✓	✓	✓
Settlement interventions	(✓)	✓	✓	✓	✓	✓	✓
Livestock offtake							
Commercial livestock offtake	✓	—	—	✓	✓	—	—
Slaughter livestock offtake for consumption	✓	—	—	—	✓	✓	—
Provision of livestock							
Replacing lost livestock assets	—	—	✓	—	—	—	✓
Building livestock assets	—	—	✓	—	—	—	✓

Annex D: Acknowledgements and contributors

Editorial Committee (LEGS Technical Advisory Group)

Andy Catley (Feinstein International Center, Friedman School of Nutrition Science and Policy, Tufts University)

Wendy Fenton (Humanitarian Policy Group, ODI)

Guido Govoni (International Committee of the Red Cross)

David Hadrill, Chair (Independent)

Vikrant Mahajan (Sphere India)

Rosanne Marchesich (FAO)

Emmanuella Olesambu (FAO)

Piers Simpkin (Independent)

Handbook revision team

Helen de Jode (Handbook revision editor)

Cathy Watson (LEGS Coordinator)

Chapter revision authors

How to Use this Handbook – Helen de Jode

Chapter 1: Introduction to LEGS – David Hadrill

Chapter 2: LEGS Principles – Andy Catley

Chapter 3: Emergency Response Planning – Cathy Watson, Polly Bodgener, David Hadrill and Andy Catley

Chapter 4: Livestock Feed – Adrian Cullis

Chapter 5: Water – Mathias Frese

Chapter 6: Veterinary Support – Andy Catley

Chapter 7: Shelter and settlement – James Kennedy

Chapter 8: Livestock offtake – Yacob Aklilu

Chapter 9: Provision of livestock – Raphael Lotira and Adrian Cullis

Annexes

Consultation process

A wide range of people, too many to be named here, provided their expertise in the production of the third edition of the LEGS Handbook. LEGS gratefully acknowledges all contributors for their valuable inputs including, among others:

- all who reviewed the online consultation version of the revised Handbook and provided detailed comments in 2022
- participants in an online survey on the 2nd edition in 2021
- the convenors and participants in five consultation workshops in India, Kenya, Mali, Nicaragua, and the Philippines on the 2nd edition of LEGS in 2020/2021
- the authors of seven discussion papers in 2020 (gender, nutrition, resilience, livestock insurance, Covid-19, institutionalisation and localisation, and the quality of veterinary pharmaceuticals) and the participants in the related webinars

Technical advice

Xavier Argoud – Emergency response

Julia Ashmore – Shelter and settlement

Suzan Bishop – Emergency response

Pauline Gitonga – Veterinary support

Kate Hart – Cash and voucher assistance

Racey Henderson – SEADS

Diane Hiscock – Inclusion

Karin de Jonge – Gender

Emma Jowett – Cash and voucher assistance

Simon Kihu – Veterinary support

Tanya Lutvey – Gender

Lucy Maarse – Gender

Genene Regassa – Feed

Abdirashid Saleh – Peri-urban response

Kebadu Simachew Belay – Feed

Editorial support

Plain English edit: Jane Lanigan/Editors4Change Ltd

Copy editing: Cordelia Lily

Indexing: Pierke Bosschieter

Design

www.truedesign.co.uk

Typesetting

River Valley Technology

LEGS Secretariat

Cathy Watson (LEGS Coordinator)

Suzan Bishop (LEGS Technical and Project Manager)

Lucy Margetts (LEGS Administration and Finance Manager)

Brianne Miers (LEGS Communications Officer)

Donors

LEGS gratefully acknowledges the US Agency for International Development Bureau for Humanitarian Assistance for financial support to revise and produce the LEGS Handbook 3rd Edition. LEGS also acknowledges the support (cash and in-kind) from the following organisations over the last four years: FAO, ICRC, ODI, PRODEL, Sphere, Sphere India, The Brooke, Tufts University, and VSF-Belgium.

Index

Page numbers in **bold** refer to figures. Entries in **bold** are LEGS technical interventions. Entries <u>underlined</u> are LEGS Principles.

A

accountability
   in emergency response planning 103
   and MEAL 66
      see also Monitoring, evaluation, accountability and learning
administrative systems 154
advocacy *see* policy/advocacy context
affordability, of veterinary support 222–223
AMR (antimicrobal resistance) 35
animal husbandry, training in 350
animal welfare 22, 40
animals *see* breeding animals, companion animals, livestock working animals
antibiotics 35
antimicrobial resistance (AMR) 35, 41
apiculture 43
aquaculture 43
assessment methods 80
assessment teams 80–81
assessments
   participatory 229
   of risks 268
   see also initial assessments
assessments/planning
   of continuity of water trucking 195
   of demand for water 190
   of disposal needs 238
   of livestock diseases 228
   of livestock feed 154–155
   of livestock offtake 315
   of livestock provision 352–353
   of livestock shelter/settlement 270–272, 282
   of local markets 272
   of veterinary support 241
   of water points 199
   of zoonotic diseases 236
at-risk groups
   climate change 56
   and community participation 53
   and livestock shelter/settlement 271
   targeting of, for livestock feed 145
   and veterinary support 231–232

B

bank accounts 121
benefit-cost analysis 103, 134–135, 343–344

benefits/challenges
   in general 92
   of feed camp emergency feeding 140–141
   of feed supply options 140–141
   of home-based emergency feeding 140
   of livelihoods-based livestock programming 48
   of livestock feed 137
   of livestock offtake 295
   of livestock provision 329
   of livestock shelter/settlement 258–259
   of MEAL 66
   of veterinary support 214–216
   of water provision 175–177
boys 62
breeding animals 135–136, 142, 323, 326–327, 329, 355
budgets 157–158, 358

C

CAHWs (community-based animal health workers) 209, 230, 232–234
calendars, seasonal 105
camps
   livestock provision to people living in 337–338
   and veterinary support 223
   see also feed camps
carcasses, condemned 314
cash and voucher assistance (CVA)
   in general 25, 119–120
   context of 86
   decision tree for **122**
   delivery mechanisms for 120
   further reading on 127
   in livestock feed 156
   in livestock offtake 311
   in livestock provision 337, 346–347
   and livestock shelters 265
   and response modalities 94
   in veterinary support 223, 244
   in water provision 173
cash money 120
checklists
   for feed provision 161–162
   for initial assessments **81**, 81–86, 106–107, 109
   for livestock offtake 315–316

Index

for livestock provision 352
for livestock shelter/settlement 282–283
for nature and impact of the emergency 83–85
for role of livestock in livelihoods 82
for situation analysis 85–86
for veterinary support 241
for water points 199
chronic/recurring emergencies 28
CHS (Core Humanitarian Standard on Quality and Accountability) 37, 39, 70–72
climate change 55–57
 further reading on 73
 impacts of 56
 and livestock 35, 55–56
 and livestock feed 144
 and livestock offtake 299
 and livestock provision 332
 and livestock shelter/settlement 262
 and veterinary support 219
 and water provision 180
clinical veterinary services
 in general 208–210
 benefits/challenges of 214–215
 design of 230–234
 examination/treatment of animals in 210
 mass medication/vaccination programmes 210
 payments options for 222–223, 233–234
 roles and responsibilities in 235
cohabitation, of livestock and humans 253, 277
community leadership
 and water provision
  see also local ownership
community participation 52–55
 active 53–55, 59
 and CHS 70
 and direct communication 54
 emergency livestock projects 52–53
 further reading on 73
 in livestock feed 143
 and livestock offtake 299, 306, 311
 and livestock provision 332
 and livestock shelter/settlement 262, 271–273
 types of 54–55
 and veterinary support 218
 and water provision 179–180
 and wider stakeholders 272–273
community shelters 275
community-based animal health workers (CAHWs) 209–210, 230, 232–234
community-based approaches 232
companion animals 44
companion standards 37

complementary programming, and livestock feed interventions 156
complex emergencies 28, 30–32, 36, **117**, 211–212
content, of LEGS Handbook 15
contingency planning
 and DRR 58–59
 for veterinary support 226
control
 of livestock diseases 240
 over livestock 62–63
 of quality 159
 of zoonotic diseases 236–237
coordinated responses 59–61
 between different interventions 61
 between livestock interventions 60
 and CHS 71–72
 further reading on 74
 in livestock feed 144
 and livestock offtake 300
 and livestock provision 333
 and livestock shelter/settlement 262
 and veterinary support 219
 and water provision 180
coordination groups
 between groups, for livestock offtake 307
coping strategies 108, 151–152
Core Humanitarian Standard on Quality and Accountability (CHS) 37, 39, 70–72
cost recovery 193, 203, 222
cost-effectiveness 212
Covid-19 pandemic 34
CPMS (Minimum Standards for Child Protection) 61
credit systems 347
crisis modifiers 58, 227, 304
CVA see cash and voucher assistance (CVA)
cyclone resistance 269

D

data see information
dead livestock 238
decision trees
 for CVA **122**
 for livestock feed 148, **149**
 for livestock offtake 301, **302**
 for livestock provision **339**
 for livestock shelter/settlement 265, **266**
 for veterinary support 224, **225**
 for water provision **183**, 184
delivery mechanisms, for CVA 94–95, 120–121
design
 of livestock provision 354
 of livestock shelters 275
 of veterinary support 230, 243

disabled persons 277
disaggregating, of information 102
disaster risk management (DRM) cycle 28–29
disaster risk reduction (DRR) 58
disposal, of dead livestock 238, 314
distribution
  of livestock feed 154, 161
  timing of, in livestock provision 348
  and water trucking 198
donkeys 361
DRR (disaster risk reduction) 58
due diligence approach 264

E

early actions
  in livestock offtake 304
  see also preparedness/early actions
early warning systems 59, 303
earthquake resistance 268
the elderly 54
emergencies
  impact of 32
  phases in **31**, 30–31
  see also under specific emergencies
Emergency Market Mapping Analysis (EMMA) 298
emergency response planning
  in general 78–79
  accountablility in 103
  analysis of technical interventions 90–95
  and CHS 70
  impact evaluation 103
  initial assessments in 79–86, 105–111
  monitoring in 103
  response identification 87–90
  response plans 96–98
  sharing of learning in 104
emergency types 28–30
EMMA (Emergency Market Mapping Analysis) 298
environmental health 35
environmental impact
  assessment of 108
  of livestock in settlements 278
  and livestock shelter/settlement 271–272
  of use of water sources 187
environmental protection 55–57
  in general 55–57
  further reading on 73
  and livestock feed 144
  and livestock offtake 299
  and livestock provision 332
  and livestock shelter/settlement 262
  and veterinary support 219
  and water provision 180

equipment, livestock related 351
equipment, livestock-related 314
euthanasia 235
evacuation 269, 278
evaluation
  in general 66
  see also Monitoring, evaluation, accountability and learning
  in emergency response planning 103
  of livestock feed 163
  of livestock offtake 317
  of livestock provision 354–357
  of livestock shelter/settlement 284
  prioritising impact 67
  of veterinary support 243
evidence-based, good practice 21
existing users of water sources 195
exit strategies, in lifestock offtake 306

F

feed see livestock feed
feed camps 91, 139, 156, 162
fire safety 278
flood impact mitigation 268
focus group discussions 105, 109, 110
food hygiene 237
food security 350–351
funding 58–59, 227
further reading
  on initial assessments 126
  on LEGS Principles 73
  on livestock feed 165–166
  on livestock offtake 319
  on livestock provision 360–361
  on livestock shelter/settlement 287
  on participatory methodologies 128
  on veterinary support 245–247
  on water provision 203

G

gender, and power dynamics 107
gender-based violence 277
gender-sensitive programming
  in general 62
  and CHS 71
  further reading on 74–75
  and livestock feed 145
  and livestock offtake 300
  and livestock provision 334
  and livestock shelter/settlement 262
  and MEAL 68
  and technical interventions 62–63
  and veterinary support 220
  and water provision 181
good practice, evidence based 21

Index

greenhouse gas emissions 35

H

herd size 358–359
housing, land and property rights (HLP) 264
HSP (Humanitarian Standards Partnership) 37, 42
human health 35–36
human rights 21
Humanitarian Charter 37–38
Humanitarian Standards Partnership (HSP) 37, 42
humanitarian-development-peace nexus 36

I

immediate benefits, by using existing animals 26, 133, 171, 291
impact assessments 66–67
impact evaluation 67, 103
impact indicators 98
impacts
  of climate change 56
  of emergencies 32
  and mitigation measures 342
  of pandemics 33, 34
  see also environmental impact
in-kind contributions 312
INEE (Minimum Standards for Education) 61
information
  on aquaculture 43
  on companion animals 44
  disaggregating of 102
  on livestock diseases/pests 43
  on livestock insurance 44
  numerical 102
  participatory methods and other 101–102
  recording and analysing of 86
  review of existing 80
  on sustainable livelihoods 40
  on technical response interventions 39
  on veterinary medicine 41
initial assessments
  additional methodologies to support 107–108
  checklists for **81**, 81–86, 106, 107, 109–110
  further reading on 126
  gender-sensitive programming and 63
  overview 79
  participatory methods for 105
  preparation for 80–81
  recording and analysing information 86
  recording template 111
insurance, of livestock 36
Internally Displaced Persons (IDPs) 147
interview methods 105, 241–242

invasive species 56
investment, in MEAL 67

K

key actions
  in general 93
  in livestock feed
    assessment /planning 154–155
    feed safety 158–159
    feeding levels 157
    preparedness 151
  in livestock offtake
    assessment/planning 304–305
    commercial livestock offtake 307
    preparedness 303
    slaughter for consumption 310
  in livestock provision
    assesment/planning 344
    defining the intervention package 344–345
    delivery systems 346
    recipient support 349
  in livestock shelter/settlement
    assessment/planning 270
    durable shelter solutions 280
    preparedness 268
    settlement 276
    shelter 273
  in veterinary support
    assessment/planning 228–229
    clinical services 230–231
    disease surveillance 239
    disposal of dead animals 238
    examination and treatment 234
    preparedness 226
    sanitation and food hygiene 237
    zoonotic diseases 236
  in water provision
    assessment/planning 188
    preparedness 186
    water points 190–192
    water trucking 194–196

L

land access 278–279
land tenure 57, 264
learning, of sharing 66–67, 104
  see also Monitoring, evaluation, accountability and learning
legal factors, and veterinary support 226–227
LEGS Principles **49**
  in general 22–23, 48–49
  and CHS commitments 70–72
  climate change *see* climate change

Index

community participation *see* community participation
coordinated responses *see* coordinated responses
gender-sensitive programming 62
livelihoods-based programming *see* livelihoods-based programming
local ownership *see* local ownership
MEAL *see* Monitoring, evaluation, accountability and learning
preparedness/early action *see* preparedness/early action
LEGS/LEGS Handbook
   and CHS 39
   differences effected by 16
   evidence-based good practice 21
   and HSP 37
   human rights basis of 38
   and Humanitarian Charter 38
   and livelihood support 20
   other resources in 16
   purpose of 14
   rights-based approach in 21
   and Sphere Handbook 37
   who should use 14–15
livelihood assets
   definition of 25
   livestock as 25
      see also livestock assets
livelihoods
   climate change impacts on 56
   definition for 20
   role of livestock in 341
   support for 20–21
   sustainable 40
   of women 23
livelihoods objectives
   in general 51
   protect key livestock assets 26, 27, 133, 171, 207, 251, 291
   rebuild key livestock assets 26, 27, 133, 171, 207, 323
   SMART objectives and 97
   use of existing livestock assets 26, 133, 171, 291
livelihoods-based programming 48–52
   benefits of 48–51
   and CSH 70
   further reading on 73
   humanitarian versus development support 51–52
   and livestock feed 143
   and livestock offtake 298
   and livestock provision 332
   and livestock shelter/settlement 261

local services and markets 51
   and MEAL 67
   and veterinary support 218
   and water provision 179
livestock
   affected communities continue to keep 155
   as livelihood assets 25
      see also livestock assets
   buyers of 308
   and climate change 35, 55–56
   dead, disposal of 238, 314
   definition of 23
   euthanasia of 235
   evacuation of 269, 278
   examination and treatment of 214
   and financial capital 50
   and fire safety 278
   and human capital 50–51
   impact of emergencies on 32
   insurance of 36
   loss of 268
   mass medication programmes for 210, 215
   mass vaccination programmes for 211, 215
   and minimum viable herd size 358
   mobility of 145–146
   mortality of 107
   movement of 57
   prices paid for 308, 311
   procurement of 346, 353
   provision of *see* livestock provision
   removal of *see* livestock offtake
   role in livelihoods of 341
   security of 263, 276
   selection of 308, 312, 344–345
   settlement of *see* livestock settlement, livestock shelter/settlement
   shelters for *see* livestock shelter/settlement, livestock shelter
   and social capital 50
   sustainable settlement of 279
   targeting of 155–156
   transport of 348
   untethering of 269–270
   waste of 57
   and water quality 189
livestock assets
   benefits of using existing 26, 133, 171, 291
   building of 328, 330, 357
   protection of 26–27, 207, 251, 291
   rebuilding of 26–27, 133, 207, 323
   replacing lost 326–329, 355
livestock buyers 308
livestock diseases/pests
   assessment of 228
   emergency guides 43

Index

epidemic 208
and livestock provision 361
minimising 147
surveillance of 213–214, 216, 239–240
transmission of 336
livestock fairs 346–347
see also livestock markets
livestock feed
in general 87, 91, 133
administrative systems in 154
assessment/planning of 154
at-risk groups in 145
benefits/challenges of 137–138, 140, 141
budgeting of 157–158
in camps 147–148
checklist for 161–162
decision tree for 148–149
external procurement of feed 153
and feed camps 139, 156
home-based 138–139
importance of 27, 134–135
and LEGS Principles 143–145
and livelihoods objectives 133
and local capacities/coping strategies 146
and local markets 147
monitoring/evaluation in 163–164
nutritional quality of 157
options for ensuring supply of 136–137
and other HSP standards 142–143
and other LEGS chapters 142–143
and pests/diseases 147
and PRIM 114, 116, 118
provision of, CVA schemes in 156
replenishment of stores of 158
safety of 158
and security 153
sourcing/distribution of 161–162
timing of 141–142
livestock insurance 44
livestock interventions
coordination between 60
other interventions and 61
see also specific interventions
livestock keepers
and alternative livelihoods 25
displaced 147–148
HLP rights for 264
impact of emergencies on 32
local environmental management 56–57
nomadic *see* pastoralists
and provision of livestock shelters 253
security of 277
sustainable settlement of 279
types of 24

see also at-risk groups, smallholder farmers
livestock markets
monitoring of 303
support for 309
understanding of 301, 307
see also livestock fairs, purchase sites
livestock offtake
in general 87, 91, 290–291
assessment/planning of 304–306
benefits/challenges of 295–296
checklists for 315–316
choice between types of 305
commercial 292–293, 307–310, 315, 317
coordination groups in 307
customary practices in 305
decision tree for 301, 302
early actions in 304
example of 125
and feed supply 136–137, 142
further reading on 319
importance of 291–292
institutional context of 306
intervention areas in 308
and LEGS Principles 298
and livelihood objectives 291
monitoring and evaluation of 317–318
and other HSP standards 298
and other LEGS chapters 298
and PRIM 115–118
procurement in 312
purchase sites 311
recipients of materials of 312
selection of animals in 305
slaughter for consumption 293–294, 310–316, 318
slaughter for disposal 294–295, 316
technical standards for 290
timing of 296, 304
transaction costs for 309
livestock production systems, large-scale 35, 56
livestock settlement
in general 257–258
assessment of 283
benefits/challenges of 259
changing needs for spaces in 281
definition of 251
overview 276–280
see also livestock shelter/settlement
livestock shelter 273–276
in general 255–256
assessment of 282
benefits/challenges of 258
and climate 255, 274, 285–286

Index

community 256
design of 274
disaster safe 273
and environmental degradation 274
and local markets 264
and security 263
transition to durable 275, 280–281
see also livestock shelter/settlement
livestock shelter/settlement
in general 27, 87, 115, 118, 251
assessment/planning of 270–272
benefits/challenges of 258–259
checklists 282–283
decision tree for 265, 266
for displaced and non-displaced populations 254–255
further reading 287
importance of 252
and LEGS Principles 261–263
livelihoods objectives 252
and livestock loss 268–269
monitoring and evaluation of 284
options for 254
and other HSP standards 261
and other LEGS chapters 260–261
and preparedness 268–270
PRIM 115, 118
technical options for 91
timing of 259–260
see also livestock settlement, livestock shelter
livestock, provision of
in general 87, 91, 323
assessment/planning of 341–343
benefits/challenges of 329–330
checklists for 352–353
decision tree for 338–339
defining the intervention package 344–345
delivery systems in 346
examples of 27
further reading on 360
impact and mitigating measures in 342
importance of 324–325
and LEGS Principles 332
and livelihood objectives 323–324
and minimum viable herd size 358–359
monitoring and evaluation of 354–357
to non-livestock keepers 343
and other HSP standards 331
and other LEGS chapters 331
and PRIM 115, 117
recipients of 335, 345–346
timing of 330–331, 348
livestock-based responses
complementary 24, 25

see also under specific livestock-based responses
livestock-keeping communities
impact of emergencies on 32
resilience 33
support for 20
local capacities 70, 88, 146, 152
local markets
in general 51
disruptions of 147
and livestock shelters 264, 272
see also livestock markets
local ownership
in general 64
and CHS 70
further reading on 75
LEGS approach for supporting 65
and livestock feed 145
and livestock offtake 301
and livestock provision 334
and livestock shelter/settlement 263
and MEAL 68
past experiences with 64
and veterinary support 220–221
and water provision 181
local services 51
localisation concept 64
logistics
of coordination with other sectors 61
of feed camps 162
of water trucking 196

M

management systems, for water sources 189, 194
mapping 105
market analysis 126–127
mass medication programmes 210–211
mass vaccination programmes 211
matrix scoring 106
MEAL *see* Monitoring, evaluation, accountability and learning
meat
distribution of 312
fresh vs dried 313
inspections of 237
Minimum Standard for Market Analysis (MISMA) 297, 298
Minimum Standards for Child Protection (CPMS) 61
Minimum Standards for Education (INEE) 61
MISMA (Minimum Standard for Market Analysis) 297, 298
MNBs (multi-nutrient blocks) 136, 153
mobile money 121, 311

Index

modalities 94
monitoring
    in general 66–67
    in emergency response planning 103
    of livestock condition 304
    of livestock feed 163
    of livestock markets 303
    of livestock offtake 310, 317
    of livestock provision 354–357
    of livestock shelter/settlement 284
    of veterinary support 243
    of water provision 201
    see also Monitoring, evaluation, accountability and learning
Monitoring, evaluation, accountability and learning 66–68
    in general 99
    benefits of 66
    and CHS 71
    and community participation 67
    in emergency response planning 99–104
    further reading on 75, 127
    and gender-sensitive programming 68
    impact indicators in 100
    implementing of 67
    and livestock feed 145
    and livestock offtake 301
    and livestock provision 334
    and livestock shelter/settlement 263
    and local ownership 68
    and other LEGS Principles 67–68
    participatory methods in 101–102
    process indicators in 100
    SMART objectives in 99–100
    and veterinary support 221
    and water provision 182
multi-nutrient blocks (MNBs) 136, 153

N

norms social/cultural 233
nutrition 107
nutritional quality 157, 161

O

older people 54
One Health approach 35–36

P

pandemics 33–34, 41
participatory assessments, in veterinary support 229
participatory mapping 241
participatory methods
    alternative approaches to 102
    further reading on 128

    for initial assessments 105–107
    and other data 101–102
Participatory Response Identification Matrix (PRIM) 70, 88, **89**, 112–118
participatory targeting 98
partnerships, in livestock offtake 308
pastoralists
    in general 24
    and CVA 95
    and livestock mobility 146
    and livestock provision 326–327, 329, 355
    vs. smallholder famers 335
pathway encroachment 278
personal security
    in livestock feed 153
    of livestock keepers 277
    and livestock offtake 306
    and livestock provision 335
    in livestock shelter and settlement 263
    and Sphere Protection Principles 69
    of veterinary personnel 221, 233
    and water provision 182, 191, 198
pests *see* livestock diseases/pests
piling, proportional 106, 110
policy/advocacy context 154, 226
pre-paid cards 120
preparedness/early action
    in general 58–59
    and CHS 70–71
    in disaster risk management cycle 28
    further reading on 74
    for future emergencies 350
    in livestock feed 144, 151–154
    in livestock offtake 300, 303–304
    in livestock provision 333
    in livestock shelter/settlement 262, 268–270
    in veterinary support 212, 219, 226–228
    in water provision 180
prices, of livestock 308, 311
PRIM (Participatory Response Identification Matrix) 70, 88, **89**, 112–118
process indicators 100
procurement, of livestock 312, 346, 353
Protection Principles 37–38
protracted emergencies 28, 31
provision
    of livestock *see* livestock provision
    of water *see under* water
public awareness, of food safety 237–238
public health impact, of livestock in settlement 279–280
public health risks, of livestock offtake 313–314
public sector veterinary functions
    in general 208–209, 211–212
    benefits/challenges of 216

Index

purchase sites, for livestock offtake 311
   see also livestock markets

Q

quality control, of feed 159
quality criteria 70–72
quality, of water 189, 196
quantitative surveys 102

R

ranking methods 105
rapid-onset emergencies 28, 30–32, **112**, **114**, 269
recipients
   of livestock
      additional support to 349–351
      checklist 353
      selection of 335–336, 336
   of meat from livestock offtake 312
redistribution systems 347–348
reporting, in disease surveillance 240
resilience 33
resilience analysis 33
resources, in LEGS Handbook 16
response identification
   overview 87
   and PRIM 88–90
response modalities, and CVA **119**
response plans
   contents of 98
   example of livestock offtake 125
   overview 96
   SMART objecties of 97
   template for 124
   theory of change for 96–97
restocking see livestock provision
rights-based approaches 21
risk assessments 268

S

safety
   of livestock feed
      checklist for 161
      and quality control 159
      and risk assessments 159
      and sanitary procedures 159–160
   see also personal security, security
samples, of PRIM 114–118
sampling 102
sanitary procedures 159, 237
scoring methods 105–106
SEADS (Standards for Supporting Crop-related Livelihoods in Emergencies) 24, 61
security
   of livestock 263, 276
   see also personal security, safety
selection
   of livestock 308, 344
   of recipients of livestock 345–346
   of recipients of meat from livestock offtake 312
self-mobilisation 55
shelter
   for humans 351
   for livestock see livestock shelter/settlement
slaughter
   for consumption 293, 310, 315–316, 318
   for disposal 316
   equipment 314
   facilities 237
   methods 313
   waste 314
slow-onset emergencies 28, 29, 31–32, **113**, **116**
smallholder farmers 24, 327–329, 335, 356
smart cards 120
SMART objectives 97, 99, 125
sourcing, of livestock feed 161
Sphere foundations 37, **49**, 69–72
Sphere Handbook 37
standards see technical standards
Standards for Supporting Crop-related Livelihoods in Emergencies (SEADS) 24, 61
stockpiling 153
supply route, for water trucking 197
sustainable livelihoods 40, 272

T

tankers 196
technical interventions **49**, 90
   decision trees in 93
   gender-sensitive programming and 62–63
   options in, outline of 92
   overview 90–91
   practical advice 39–40
   standards and guidelines in 93
   summary of 91
   timing tables in 92–93
technical standards
   for livestock provision **322**
   for livestock feed **132**
   for livestock shelter/settlement **250**
   overview **129**
   for veterinary support **206**
   for water **170**
   see also under specific subjects of standards
templates, for PRIM 112–113

Index

theory of change 96–97
timelines, historical 105
timing
  in general 92
  of combined interventions **375**
  in livestock feed 141
  of livestock offtake 296–297, 304
  of livestock provision 330–331, 348
  of livestock shelter/settlement 259
  of mass vaccination programmes 211
  of veterinary support 216–217
  of water provision 177
training, in animal husbandry 350
transaction costs, for livestock offtake 309
transport planning, in livestock provision 348–349
tsunami impact mitigation 269

U
urban and peri-urban 257–258

V
vaccines 211
Venn diagrams 105
veterinarians 234–235
veterinary investigations 239
veterinary medicines
  information on 41
  procurement and storage of 232
  protection of 233
  provision of 244
  supply, quality and storage 228
veterinary paraprofessionals 209, 222, 230, 234
veterinary personnel 221, 232–233
veterinary public health 36, 212–213, 216
veterinary service providers 227, 229–230
veterinary support
  in general 27, 57, 87, 91, 114–115, 206–207, 345
  and access to communities 221
  assessment/planning in 228–230
  benefits/challenges of 214–216
  checklist for 241–242
  clinical veterinary services 209–211
  decision tree for 224–225
  design of 227, 243
  examination and treatment in 210, 234–235
  flexible funding in 227
  further reading 245–247
  importance of 208
  and LEGS Principles 218–221
  and livelihoods objectives 207–208
  long-term 350
  monitoring and evaluation of 243–244
  and other HSP standards 217

  and other LEGS chapters 217–218
  and preparedness 212, 226
  preventive 349
  prioritising of 212
  public sector veterinary functions 211–214
  service providers 227
  timing of 216–217
  and veterinary public health 212
veterinary training 244
visualisation methods 105
volcano impact 269
vouchers
  see also CVA

W
water
  assessment of demand for 190
  conflicts over 182
  further reading 203
  and livelihoods objectives 171
  and personal security 182, 191
  provision of
    in general 27, 87, 91, 170–171, 173
    adequacy of 190–191
    assessment/planning in 188–189
    benefits/challenges of 175–177
    decision tree for 183, 184
    environmental impact of 187
    importance of 172–173
    and LEGS Principles 179–182
    monitoring/evaluation of 201
    and other HSP standards 178–179
    and other LEGS chapters 178
    and preparedness 186–187
    and PRIM 114, 116, 118
    timing of 177–178
    through water trucking see water trucking
  quality of 189, 196
  sources of
    analysis of options using existing 188
    contamination of 189
    development of 187
    for human consumption 189, 195–196
    management systems for 186–187, 189
    mapping of existing 186
    water points 190–191, 194
water points
  checklists for 199–200
  and cost recovery 193, 203
  location of 190
  management structures for 194, 202
  rehabilitation/establishment of 191–194
    in general 173–174
    benefits/challenges 175–176

Index

    need for 192–193
    suitability for 192
    technical feasibility of 193
  responsibilities for 193–194
water trucking
  in general 174–175
  as short-term measure 194–195
  assessment of the continuity of 195
  benefits/challenges of 177
  distribution points 198
  logistics and distribution for 196–198
  maintenance and fuel supplies 197
  to mobile livestock 198
  sources and quality 194–196
  sources used for 195
  and staffing 197
  subsidising of 198
  and supply routes 197
working animals 23–24
  see also livestock

Z

zoonotic diseases 33–34, 208, 212–213, 236–237, 277